Potassium
Argon Dating

Compiled by

O. A. Schaeffer and

J. Zähringer

Springer-Verlag New York Inc.

ISBN 978-3-642-87897-8   ISBN 978-3-642-87895-4 (eBook)
DOI 10.1007/978-3-642-87895-4

© by Springer-Verlag Berlin · Heidelberg 1966
Library of Congress Catalog Card Number 66-15714
Title No. 1340

Softcover reprint of the hardcover 1st edition 1966

List of Contributors

Asst. Prof. R. L. ARMSTRONG, Geological Department, Yale University, New Haven, Conn., U.S.A.

Prof. G. H. CURTIS, University of California, Berkeley, Cal., U.S.A.

Dr. H. FECHTIG, Max-Planck-Institut für Kernphysik, Heidelberg, Germany

Prof. F. G. HOUTERMANS, Physikalisches Institut der Universität Bern, Switzerland

Prof. P. M. HURLEY, Department of Geology and Geophysics, Massachusetts Institute of Technology, Cambridge, Mass., U.S.A.

Dr. S. KALBITZER, Max-Planck-Institut für Kernphysik, Heidelberg, Germany

Dr. T. KIRSTEN, Max-Planck-Institut für Kernphysik, Heidelberg, Germany

Dr. D. KRANKOWSKY, Max-Planck-Institut für Kernphysik, Heidelberg, Germany

Dr. O. MÜLLER, Max-Planck-Institut für Kernphysik, Heidelberg, Germany

Prof. O. A. SCHAEFFER, Chairman of Department of Earth and Space Sciences, State University of New York, Stony Brook, L. I., N. Y. U.S.A.

Prof. G. W. WETHERILL, Institute of Geophysics and Planetary Physics, University of California, Los Angeles, Cal., U.S.A.

Dr. J. ZÄHRINGER, Member of Directorate, Max-Planck-Institut für Kernphysik, Heidelberg, Germany

Wolfgang Gentner

This book is respectfully dedicated to WOLFGANG GENTNER on his sixtieth birthday. Although most of Gentner's scientific career is concerned with nuclear physics, he has always maintained an active interest in the applications of nuclear physics to other areas. The first paper he published was on a biological subject: „Die Strahlungsreaktion des Eiweißes und die Erythemwirkung". One of his latest publications is entitled „Das Rätseln um die Herkunft der Tektite", concerned among other things with the application of potassium argon dating to the origin of tektites.

WOLFGANG GENTNER was born on July 23, 1906 in Frankfurt am Main. He obtained his doctor's degree under Professor F. DESSAUER in 1930 from the University of Frankfurt am Main. The years from 1932 to 1935 he spent as a guest at the Radium Institute of the University of Paris. At this time the institute was directed by Madame PIERRE CURIE. From 1935 to 1946 he served as a scientific assistant to WALTER BOTHE at the Kaiser Wilhelm Institute in Heidelberg.

His early work for his doctor's degree and at the Radium Institute was concerned with physical and biological effects of gamma rays. This work was continued during his collaboration with WALTER BOTHE and resulted in pioneering work on the nuclear photo effect. During these years at Heidelberg he was also responsible for the construction of a cyclotron. During the winter of 1938–1939 he spent several months as a Helmholtz Society scholar at the Radiation Laboratory of the University of California at Berkeley.

In 1946 he went to the University of Freiburg i. Br. Here in addition to nuclear physics he became interested in researches of age determinations mainly concerned with the potassium argon method. These papers are among the earliest measurements in this field of dating. His stay at Freiburg was interrupted by the appointment as a director of the Synchro Cyclotron group at CERN, Geneva. Under his able leadership the Synchro Cyclotron was built. In 1958 he left Geneva to become director of the newly founded Max-Planck-Institut für Kernphysik in Heidelberg, which position he still holds. Under his leadership the Max-Planck-Institut at Heidelberg grew very rapidly and has become a pioneer in the field of nuclear physics, as well as its application to meteorites and geological problems. The vistas of the Institute are presently being expanded to include researches in the areas of space physics.

Gentner has received several outstanding international awards which include l'Officier de la Légion d'Honneur and the Honorary Fellowship at the Weizmann Institute of Science.

It is difficult in a few short sentences to try to get a feeling for the keen scientific interest and insight as well as the warmth of his personality. In spite of pressing administrative duties he still spends much time in the laboratory discussing the latest scientific results and arguing about interpretations of newly found data. His cordiality and generosity has led to an esprit de corps which is a privilege to those who work around him.

Preface

Perhaps no dating method has the wide range of applicability as does the potassium argon dating method from either consideration of the ranges of ages which can be dated or the availability of suitable material to date. Minerals as young as tens of thousands of years to minerals billions of years old have been successfully dated. Many minerals retain for times of the order of billions of years the daughter, Ar^{40}, and many minerals contain as a component K^{40} the parent element, potassium being a common element in the earth's crust. As a result, most rock contains at least one mineral which can be successfully dated by the potassium argon method.

Even though this method has been applied for over fifteen years, there is as yet no work which summarizes the experimental techniques and the results available. The sixtieth birthday of W. GENTNER, one of the pioneers in this field of research, is a suitable time to present such a summary. The present work is the combined efforts of several authors which it is hoped will go a long way towards filling this present gap. The work contains several chapters on experimental techniques. These chapters should serve on the one hand as a review of the present state of the art and on the other hand as a useful guide to investigators who are either presently engaged in potassium argon studies or who are presently contemplating initiating such studies. These chapters have been written in such a way that many of the pitfalls of the method are clearly delineated. In addition to the experimental techniques, there are chapters which summarize the results in several selected fields of research. This coverage is not intended to be complete but rather to illustrate with several examples some of the triumphs of potassium argon dating in the solving of problems of the earth and space sciences.

The book also contains one of the last scientific contributions of the late F. G. HOUTERMANS, a historical summary of potassium argon dating. This chapter illustrates clearly the scientific insight and the jovial sense of humor which so characterized the life of FRITZ HOUTERMANS.

The bibliography, although extensive, is certainly not a complete coverage of the literature of potassium argon dating. Undoubtedly, a number of important papers have been omitted from this list. This is solely due to the subjects chosen to illustrate the potassium argon method. We would like to acknowledge the cooperation of the authors in making their manuscripts available so that this book can be published on a relatively short time scale.

April 1966

O. A. SCHAEFFER, Stony Brook, L. I., N. Y.   J. ZÄHRINGER, Heidelberg

Table of Contents

History of the K/Ar-Method of Geochronology

By

F. G. HOUTERMANS

The history of the K/Ar-method for absolute dating of minerals and rocks is full of surprises and good guesses. The β-activity of potassium was discovered, together with that of rubidium, by J. J. THOMSON as early as 1905. It was confirmed by a considerable number of authors (for the early literature of potassium β-activity cf. MEYER and SCHWEID-LER (1927)). The γ-activity of potassium was discovered by KOHLHÖRSTER in 1928 and studied by him in potassium-bearing salt mines. In 1935, KLEMPERER and, independently, NEWMAN and WALKE (1935), ascribed, from reasons of isotope systematics, the activity of potassium to a — then unknown — rare isotope K^{40}. This was the first good guess. In 1935, A. O. NIER actually discovered this isotope and found its abundance to be $1.19 \cdot 10^{-4}$ of the total K. SMYTHE and HEMMENDINGER (1937) found that the β^--activity of K is actually due to K^{40}.

The second good guess is due to C. F. v. WEIZSÄCKER (1937). Everybody occupied with questions of the abundance of nuclides knew that there exists an excess of Ar^{40} in the earth's atmosphere by about a factor $300-1000$. All other chemically analoguous elements show a steady decline in terrestrial and cosmochemical abundance with increasing atomic weight. The heavier rare gases also follow this trend, with the exception of the abnormally high abundance of Ar^{40} in the earth's atmosphere. C. F. v. WEIZSÄCKER concluded — only from the Ar^{40} abundance evidence —:

1. The total activity of K is due to K^{40}.

2. Since it was known that potassium does not emit positrons, at least less than a few percent of its β^--emission, he assumed the existence of a process — entirely unknown experimentally at the time — namely electron capture by the nucleus, preferably from the K-shell with the emission of monochromatic neutrinos. This process had been considered before, purely theoretically, by CHR. MØLLER (1937).

On the basis of these conclusions v. WEIZSÄCKER suggested:

1. The measurement of Ar^{40} in minerals and rocks as a geochronological method of dating,

2. The entire excess of Ar^{40} in the earth's atmosphere as due to K^{40} decay by electron capture.

He estimated the probability of electron capture to be inferior, but comparable to that of β^--decay. During the next decade some properties of the K^{40} decay were established. O. HIRZEL and H. WÄFFLER (1946) measured the energy of the γ-radiation as 1.54 MeV, a value that was confirmed by H. A. MEYER et al. (1947) who also undertook coincidence experiments between β^-- and γ-activity, with a definitely negative result. The upper limit of the β^--spectrum was determined by DZELEPOW et al. (1946) to be 1.35 MeV, a value quite in agreement with more modern measurements. T. R. ROBERTS and A. O. NIER (1950) have measured the mass difference Ca^{40}-Ar^{40} with a precision double focussing mass spectrometer, finding $3.2 \pm 0.8 \cdot 10^{-4}$ atomic mass units against a value of $2.7 \pm 2.1 \cdot 10^{-4}$ atomic mass units obtained by SAILOR (1949) from disintegration data. This proved definitely that the excited state of Ar^{40} giving rise to its γ-emission lies only 0.17 ± 0.13 MeV below the ground state of K^{40} and, therefore, cannot be reached by β^+-emission.

In 1947 a paper by two Swiss authors, E. BLEULER and M. GABRIEL caused utter confusion in the field of potassium decay. These authors claimed to have found experimental evidence for the electron capture of K^{40} by detection of the K_α-emission of potassium and its resulting Auger electrons. God only knows what they measured and how they obtained their results. They gave the electron capture rate as 1.9 times the rate of β^--decay, i.e. about 39.5 times higher than it is known today to be true. In addition they gave the decay constant λ_{β^-} by a factor 2.1 too high. This resulted in a total half life for K^{40} as low as $2.4 \cdot 10^8$a instead of $1.31 \cdot 10^9$a as it is known today. Such a result would have had very serious geological consequences. It would have meant that, at the time of the formation of the earth's crust and most meteorites, 98% of the potassium would have consisted of K^{40}! This would have been in strict contradiction to the well known rule of isotope systematics, namely that isotopes with an odd number of protons and an odd number of neutrons are very rare, at least above N^{14}.

The author of this article used to refer to the paper of BLEULER and GABRIEL in a number of seminars he gave in England, Germany and Italy in the late forties — with his usual considerable exaggeration — as to the paper of the red hot saurians. GOODMAN and EVANS (1941) calculated the heat output from acidic igneous rocks for an average uranium, thorium and potassium content as $2.2 \pm 0.2 \cdot 10^{-6}$ cal/g year for uranium, $2.6 \pm 0.4 \cdot 10^{-6}$ cal/g year for thorium and $0.14 \cdot 10^{-6}$ cal/g year for potassium. The correct value for potassium, calculated with the half life $1.31 \cdot 10^9$a accepted today, is $0.23 \cdot 10^{-6}$ cal/g year. But, with the figures given by BLEULER and GABRIEL, the heat output would have amounted to $1.1 \pm 0.2 \cdot 10^{-6}$ cal/g year, as has been pointed out by GLEDITSCH and GRAF (1947).This value was corrected in a later paper (GRAF, 1948) to $0.6 \pm 0.1 \cdot 10^{-6}$ cal/g year, but was still too high, because they believed in BLEULER and GABRIEL. Moreover, according to these authors, the heat production from acidic igneous rock at a time $2.4 \cdot 10^9$ years ago, i.e. 10 times their half life, would have amounted to 1024 times the present value, or 10^{-3} cal/g year. At the time of the formation of the earth's crust, $4.5 \cdot 10^9$ years ago, the heat

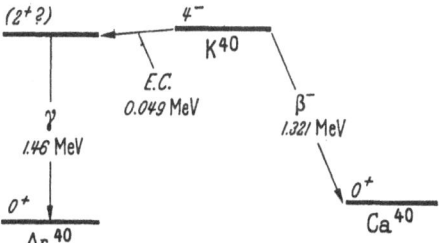

Fig. 1. Decay scheme of K^{40} according to Nuclear Data Sheets

production should have amounted to $5 \cdot 10^5$ times the present value or nearly 1 cal/g year! BIRCH has pointed out that the heat production following from BLEULER and GABRIEL would most probably have prevented solidification of rocks at an early date (1947). Therefore, the red hot saurians.

The next good guess is due to H. E. SUESS (1948). This author is noted for the fact that he comes to right conclusions on very scanty evidence or no evidence at all. In this case, he had searched for argon, together with HARTECK (1947), in two samples of sylvinite and one of carnallite, and they had found none, i.e. $\leqq 2 \cdot 10^{-4}$ cc of Ar/g, while from the decay figures of BLEULER and GABRIEL a content of $1.5 \cdot 10^{-2}$ cc/g should have followed. This could have been due — and probably was — to a loss of argon, because of re-crystallization in recent times. Together with some considerations on isotopic systematics at the time of formation of the elements involving the abundance of Lu^{176}, it was sufficient for SUESS to suggest the right decay scheme for K^{40}, given in Fig. 1. From the good agreement between the electron capture rate determined from geological evidence and the γ-emission rate determined by counting, it can be concluded that the electron capture rate directly to the ground level of Ar^{40} is probably less than 0.3% of the total decay rate.

A large number of papers has been published mainly between 1948 and 1958 on the β^- and γ-decay rate of K^{40} and on the branching ratio. These papers used direct counting measurements as well as the geological method of determining the production of Ar^{40} in minerals of known age.

Table 1. *Determinations of the specific gamma and beta activity of natural potassium*

Investigators	Activity	
	γ/g sec	β/g sec
GLEDITSCH, and GRAF (1947)	3.6 $\pm$ 0.8	
HESS, and ROLL (1948)	2.6	
AHRENS, and EVANS (1948)	3.42 $\pm$ 0.07	
GRAF (1948)		26.8 $\pm$ 1.2
BORST, and FLOYD (1948)		25.3 $\pm$ 3
HIRZEL, and WÄFFLER (1948)		33.5
GERLING, and TITOW (1949)	3.52	28.4
STOUT (1949)		30.6 $\pm$ 2.0
SAWYER, and WIEDENBECK (1949)	2.88 $\pm$ 0.3	
FAUST (1950)	3.6 $\pm$ 0.4	31.2 $\pm$ 3.0
GRAF (1950)	3.4 $\pm$ 0.5	
HOUTERMANS, HAXEL, and HEINTZE (1950)	3.1 $\pm$ 0.3	27.1 $\pm$ 1.5
INGHRAM, BROWN, PATTERSON, and HESS (1950)	3.58	28.4
SAWYER, and WIEDENBECK (1950)		28.3 $\pm$ 1.0
SMALLER, MAY, and FREEDMAN (1950)		22.5 $\pm$ 0.7
SPIERS (1950)	2.97	30.5
DELANEY (1951)		32.0 $\pm$ 3
GOOD (1951)		27.1 $\pm$ 0.6
BURCH (1953)	3.37 $\pm$ 0.09	
WASSERBURG, and HAYDEN (1954b)	3.66	28.2
BACKENSTOSS, and GOEBEL (1955)	3.50 $\pm$ 0.14	
GERLING, JASCHENKO, ERMOLIN, and BARKAN (1959)	3.58	28.4
KONO (1955)		29.2 $\pm$ 1.0
SUTTLE, and LIBBY (1955)	2.96 $\pm$ 0.3	29.6 $\pm$ 0.7
MCNAIR, GLOVER, and WILSON (1956)	3.33 $\pm$ 0.15	
WETHERILL, TILTON, DAVIS, and ALDRICH (1956a)	3.21	27.4
WETHERILL, WASSERBURG, ALDRICH, TILTON, and HAYDEN (1956d)	3.24 $\pm$ 0.15	
GERLING, JASCHENKO, and ERMOLIN (1957)	3.47	28.2
WETHERILL (1957)	3.39 $\pm$ 0.12	
GERLING (1958)	3.17	27.3
KELLY, BEARD, and PETERS (1959)		27.3 $\pm$ 0.5
EGELKRAUT, and LEUTZ (1960)	3.35 $\pm$ 0.15	28.3 $\pm$ 0.15
SAHA, and GUPTA (1961)	3.22 $\pm$ 0.15	28.8 $\pm$ 0.9
FLEISHMAN, and GLAZUNOV (1962)		27.17 $\pm$ 0.09
LEUTZ, SCHULZ, and WENNINGER (1965)	3.25 $\pm$ 0.07	28.26 $\pm$ 0.05

It is impossible here to treat every single one of these numerous papers and we shall only give a survey of them in Table 1. This table does not claim to be complete, because some quotations may have escaped the notice of the author. We shall choose as the best values those of WETHERILL et al. (1956a), obtained by the geological method, but in satisfactory agreement with counting experiments (ENDT and KLUYVER, 1954) namely $\lambda_c = 0.577 \pm 0.026 \cdot 10^{-10} a^{-1}$ and $\lambda_\beta = 4.72 \pm 0.05 \times 10^{-10} a^{-1}$ leading to a branching ratio $\lambda_c/\lambda = 11.7 \pm 1.5\%$ and a total half life of $1.31 \pm 0.02 \cdot 10^9$ a.

The next step was taken by ALDRICH and NIER (1948). They extracted argon from 4 potassium containing minerals of different age. They found an excess of Ar^{40} over Ar^{36} in argon extracted from the potassium minerals in comparison to argon from atmospheric air. Thus, they proved definitely for the first time that Ar^{40} is produced from the decay of K^{40}. At that time the extraction methods for argon were not sufficiently developed, nor was the argon half life known well enough to allow them to use their values for exact age determinations.

The next good guess is due to GENTNER and his school and this concerned the right sampling. Sampling is of extreme importance in geochronology, and it took years to choose the right samples for the K/Ar-method. SUESS had consulted chemists and mineralogists who were specialists for salt mines, but they gave him samples which later turned out to have been re-crystallized in recent times. GENTNER was well advised to choose a sample, a sylvinite, from Buggingen in the Upper Rhine Valley, belonging geologically to the Oligocene of the Upper Rhine Valley. He assumed — disregarding BLEULER and GABRIEL — according to SPIERS (1950) a rate of electron capture identical with the γ-emission rate of 3.0 γ/sec g K. The age GENTNER and coworkers obtained for the Buggingen deposit was $20 \cdot 10^6$a (PAHL et al., 1950; SMITS and GENTNER, 1950). In several later papers (GENTNER et al. 1953a; GENTNER et al., 1954b) they corrected their results for diffusion of argon in dependence of grain size and made also comparisons with the result of the He/U-method, correcting of course also for He diffusion. Their results were $25 ^{+5}_{-3}$ million years. This was the first reliable age determination with the K/Ar-method. Its success was not only due to the right choice of sample, but also to the development of a reliable and complete method of argon extraction, which, in later papers, was extended to minerals non soluble in water (GENTNER and KLEY, 1955).

Since this pioneer work of the GENTNER group the K/Ar-method became a widely used tool of geochronology. This is not the place to quote all the many hundreds, if not thousands, of age determinations by this method. One has learned that the argon retention is highest in micas, especially biotites, somewhat inferior in potassium feldspars, amphibolites, etc., but that it can be applied to a wide range of potassium containing minerals. For rock dating, very careful biotite separation is necessary. It is very fortunate that biotites also contain rubidium, so that a comparison between the K/Ar- and the Rb/Sr-method on the same mineral becomes possible. Ar and Sr are very different chemically, so that from the agreement or disagreement between the two ages important conclusions, not only with regard to the time of formation or metamorphism, but also on the conditions prevailing during these processes can be drawn.

K/Ar dating of tektites has added important clues to the extremely controversial question of their origin. GENTNER and coworkers (GENTNER et al., 1963; GENTNER et al., 1964) were able to show that the K/Ar age of the moldavites is the same as the K/Ar age of the Nördlinger Ries impact crater ($14.8 \cdot 10^6$a) and also that the ages of the Bosumtwi crater and the Ivory Cost tektites are identical ($1.3 \cdot 10^6$a). These results imply very strongly a genetic relationship between tektites and the corresponding impact craters.

GERLING and PAVLOVA (1951) were the first to apply the K/Ar-method to the dating of stone meteorites as early as 1951. Their results lacked the precision reached later by other investigators (WASSERBURG and HAYDEN, 1955; GEISS and HESS, 1958) but they nevertheless showed already that stone meteorites are much older than any terrestrial rocks. A large number of meteorites has been dated since then, many of them by a group working under GENTNER's direction (KIRSTEN et al., 1963; ZÄHRINGER, 1965). Most K/Ar ages of stone meteorites are approximately $4.5 \cdot 10^9$a, a figure in good agreement with the Rb/Sr and U/Pb ages of meteorites and also identical with the age of the earth. A few stone meteorites, however, are much younger, having K/Ar ages on the order of $6 \cdot 10^8$ a. Results such as these have given a great impetus to the whole field of meteorite research and have led to many speculations and theories on the origin and history of meteorites and the solar system.

Dating of iron meteorites with the K/Ar-method has also been attempted. This is an extremely difficult task since the potassium content of iron meteorites is extremely low, only a few ppb, and contamination problems become quite serious. The ages obtained, ranging between $6 \cdot 10^9$a and $10 \cdot 10^9$a(STOENNER and ZÄHRINGER, 1958) pose difficult questions and are thus still somewhat controversial. An age of $9 \cdot 10^9$ years for iron meteorites would require that the uranium in the meteorite consisted originally of more than 92.5% U^{235} (WEFELMEIER, cf. F. G. HOUTERMANS, 1947), quite in contradiction with our present ideas on the formation of the elements. Better experimental techniques and a more profound understanding of the basic assumptions and limitations of the K/Ar-method may help in solving the iron meteorite age puzzle and many other yet unanswered questions.

Determination of Radiogenic Argon

By

T. Kirsten

I. Introduction

The principal advantage of the K-Ar-method is seen in its wide range of application. Potassium is one of the elements which occur in great abundance in the earths crust. It is present in nearly every mineral, either as a principal constituent or as a trace element.

Since the decay product Ar^{40} is an inactive gas, the fractionation between mother and daughter product is extremely large in rock forming processes. The previously produced Ar^{40} has escaped to a large extent and the initial concentration of Ar^{40} in a crystallizing rock or mineral is practically zero. Such a complete separation is one of the principal prerequisites to any age determination method using radioactive isotopes. Furthermore, small quantities of argon are rather easy to detect, because argon belongs to the rare gases. Since very young samples as well as minerals poor in potassium can be dated, the K-Ar-method covers nearly the whole range of the time scale. The disadvantage of the method lies in the high diffusion rate of gases. Rocks which have been exposed to elevated temperatures after their primary formation lose a great deal of their argon during their geological history. The K-Ar-ages obtained in this case are lower limits. However, it is possible to obtain additional information about geological events in this way, especially when ages are available which were determined by other methods. It is to be expected that the K-Ar-method will more frequently be used for studies of the thermal history of the crust of our earth. In the following, we shall confine to the experimental methods of argon determination.

For a quantitative analysis, first the gases have to be completely extracted from the sample. Secondly, a purification by gettering all active components and a separation of argon from other noble gases is necessary. Finally, the amount of argon and the contribution of atmospheric contamination have to be determined by mass spectrometric analysis.

Atmospheric argon normally consists of three stable isotopes Ar^{36}, Ar^{38} and Ar^{40} with abundances of 0.337%, 0.063% and 99.6% respectively. Radiogenic argon has the mass number 40. The high abundance of Ar^{40} in the atmosphere is due to argon released from old crust material.

The difficulties in the determination of radiogenic Ar^{40} are as follows: The classical methods are usually not sensitive enough to detect amounts of argon in the order of 10^{-6} cm³ STP. This difficulty could be overcome by increasing the sample weight.

Since this would cause unreasonable costs for the extraction equipment, a more sensitive detection method is desirable. A further difficulty is that the volumetric method alone is not sufficient to determine the radiogenic argon. All samples have been exposed to the atmosphere and the residual gas in the system also contributes atmospheric argon. The atmosphere contains 0.89% argon and small amounts are always adsorbed by the sample. In young samples the amount of atmospheric argon can be in the same order of magnitude as the radiogenic argon. However, there is a possibility to separate the contaminating argon from the radiogenic argon by determining the content of Ar^{36}. The ratio of Ar^{40}/Ar^{36} in atmospheric argon is 295.6, therefore the radiogenic argon is

$$Ar^{40}_{rad} = Ar^{40}_{total} - 295.6 \times Ar^{36}.$$

This makes it necessary for potassium argon age determinations to measure both the isotopic composition and the amount of the argon present in the sample.

This is the reason why almost all laboratories use mass spectrometry for noble gas analyses. Sometimes the neutron activation technique is also applied or even necessary. Very often special equipment to achieve ultra high vacuum must be used to keep the amount of atmospheric argon in the residual gas below the amount of radiogenic argon.

II. Extraction and separation of argon
A. General remarks

The extraction procedure has to satisfy the following conditions:

1. Argon has to be extracted quantitatively,
2. all argon losses have to be prevented,
3. contamination of atmospheric argon should be small compared with the argon in the sample.

The extraction of the gases is accomplished by decomposition in a vacuum-system (Paneth and Peters, 1928). Presently two different procedures are used:

1. The sample is melted without flux at temperatures up to 2000° C (heat extraction).

2. The sample is decomposed chemically by means of a flux at a temperature below 1000° C (flux-method).

B. Heat extraction method

The apparatus used for the heat extraction method consists of two main parts:

1. A furnace, containing a crucible, located within a vacuum-system whose walls are cooled with water.

2. A vacuum-line with provision for the storage of the samples and for the rough purification during the degassing process.

a) Furnaces

Since high temperatures are necssary for the melting process and no gas can be lost, only a heater located within the vacuum-system is applicable. A water-cooling system surrounds the apparatus of glass.

The construction of the furnace therefore is decided by the heating procedure. For resistance heating either metal or glass can be used as materials of construction; however, for an inductive heating only non conducting materials such as vycor for instance can be used.

Before discussing the melting processes, some remarks should be made concerning the form' of the melting material and its dependence on the construction of the apparatus. Usually, the furnaces are constructed for sample weights of a few grams. The samples are packed in aluminium foils which are usually free of contaminating gases. When micas are melted, it is preferable to pack them in higher melting foils such as platinum or molybdenum in order to prevent parts of the mica from spraying out of the crucible. The best material for this purpose is platinum, because it does melt and the crucible can be used again for other samples while on the contrary molybdenum foils do not melt below 2000°C; they keep their shape and thus make it impossible to melt additional samples within the same run.

Metal furnace for resistance heating

A resistance heated furnace is shown in Fig. 1. It consists of a double wall cylinder made of stainless steel, 25 cm high. On the bottom and top of this cylinder are flanges which can be easily removed. All parts are provided with water cooling.

Two connections for a platinum/platinum-rhodium thermocouple and two massive rods which serve as current conductors are introduced into the vacuum system and sealed with a metal glass joint. A tungsten wire one mm in diameter is coiled around an aluminiumoxide support and is fixed to the ends of the rods with screws. Another piece of aluminium-oxide is necessary to support the bottom of the crucible and the heater coil which cannot support its own weight when tested. The holder of the coil is lined with two thin laminas of molybdenum before the molybdenum crucible is put in. These molybdenum laminas are to prevent a sinter process between the crucible and the coil holder and help to prolong its lifetime. At the end of the run the molybdenum lining can be pulled out of the furnace together with the crucible. It is possible to wind the heating wire directly on an aluminiumoxide crucible, but it is inexpedient because the heating wire has to be renewed very often.

Silicate melt destroys the crucible by exchange reactions between aluminiumoxide and magnesiumoxide in a relatively short time.

The samples can be changed just by opening the bottom flange. The leads for the current and the cooling-water are flexible and must not be disconnected from the bottom. A plain glass window is located in the

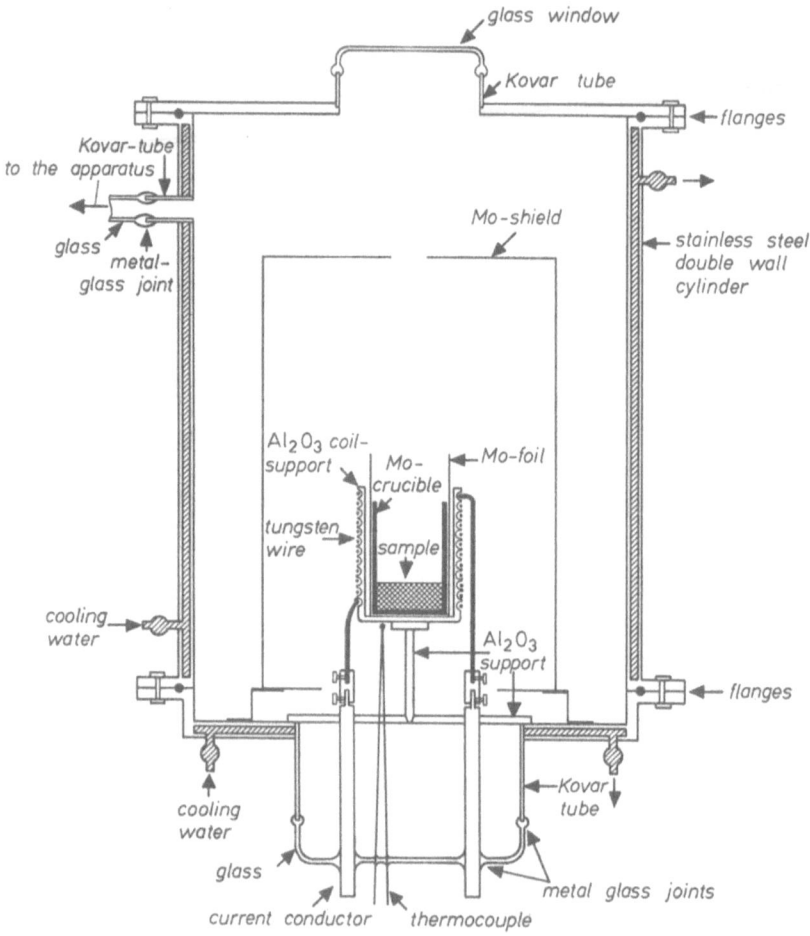

Fig. 1. Furnace with resistance heating

middle of the furnace cover to permit observations during the melting process as well as for a pyrometric determination of the temperature. The cover also contains a Kovar-tube for a connection to the vacuum-line. The heating current ranges up to fifty ampere. The temperature in the crucible can be estimated by measuring simultaneously both the heating current and the heating voltage and knowing the dependence of temperature on the specific resistance, $\varrho(T)$, of the tungsten wire. Temperatures up to 1800° C can be reached with such a furnace.

Details of the arrangement may be varied. One could, for instance, use metal seals and bake the whole furnace at about 300° C for ultra high vacuum conditions. This could be accomplished by adding another heater outside of the furnace. The ceramic parts may consist of magnesium-oxide. Tantalum can be used as a material for crucibles instead of molybdenum.

A short horizontal glass-tube in which several samples can be stored is fastened on the inside of the furnace cover, where they can be moved magnetically (with the aid of a metallic piece) into the crucible which now can be baked out before and after every melting.

If the samples contain large amounts of active gases (especially oxygen) the heating coil will be quickly destroyed. This disadvantage of the furnace limits its application.

Inductively heated glass or quartz furnaces

The most suitable apparatus, which is compatible with modern high vacuum technology, consists entirely of glass, has no greased joints or flanges and can be baked out for complete degassing. For changing samples it is necessary to cut off the end of the apparatus, clean the furnace and then fuse it together again. Any reactions between the heater and the active gases from the samples could be avoided by using an inductive heater from outside of the vacuum system. The materials for the apparatus are either special glasses (e.g. Pyrex-glass) or vycor. It is not recommended to use quartz furnaces if helium is also to be measured because of the rather high permeability of quartz for helium even at low temperatures. Fig. 2 shows a furnace of Pyrex-glass suitable for inductive heating.

A quartz container with a molybdenum or tantalum crucible on a shaft is placed inside the water cooled furnace. A separate dish of quartz with a getter material for the rough purification procedure is also placed inside the furnace (see below). The sample is melted by inductive heating of the metallic crucible with an induction coil around the furnace and at the same height as the crucible. Transfer of energy between the coil and the crucible depends on their mutual arrangement and the form of the coil (which is dependent on the frequency). Using a 1000 kc-generator and high voltage output, the best arrangement is reached with a coil having 6 to 8 turns, the diameter of which just fits the outer wall of the furnace (with the water jacket). The coils are made out of hollow copper tubes with a rectangular cross-section for the cooling-water. For a

400 kc-generator it is preferably to take a thick coil with one or two rectangular windings and a cross-section of about one cm². With pressures between 10^{-5} and 10^{-1} mm Hg high frequency discharges may occur and it is possible that argon will be included in a metallic mirror formed along the cool walls. These high frequency-discharges preferably appear at high frequencies, but can be prevented almost completely by using a 300 kc-generator. The frequency of the generator should be chosen as low as possible (around a few hundred kc).

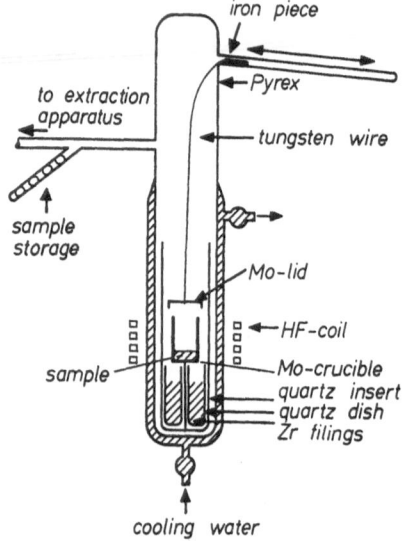

Fig. 2. Induction heated furnace

To heat one of the commonly used crucibles (3 cm high), a ten kw-generator has sufficient energy to reach temperatures up to 1800° C. Most of the silicates are eutectica and have lower melting points. For special problems (e.g. to melt pure olivine or tektites) 2000° C might be necessary to get a complete degassing. Therefore the energy of the generator has to be at least ten kw. It is better to use a twenty kw generator placed as close as possible to the furnace (in order to avoid loss of energy by long cables).

The crucible may be covered with a molybdenum lid in order to prevent small grains from spraying out when heated. It can be moved magnetically by using a tungsten wire fixed at a lateral wall tube. To determine temperatures a thermocouple can be placed within the furnace.

Finally, an additional method for melting metallic samples (iron-meteorites, ores) without any crucible should be mentioned. This is very important to decrease the blank run if extremely small amounts of argon are to be determined. The water cooled coil is brought into the vacuum system and the sample suspended, by a tungsten thread, is allowed to hang inside of the coil. As the metal melts it drops into a quartz dish at the bottom of the furnace.

b) Apparatus and melting process

The apparatus is connected with the furnace by a glass tube and should be made of a special glass capable to withstand comparatively high temperatures without showing early crystallization phenomena. The permeability of this glass for helium should be low and the line should be free

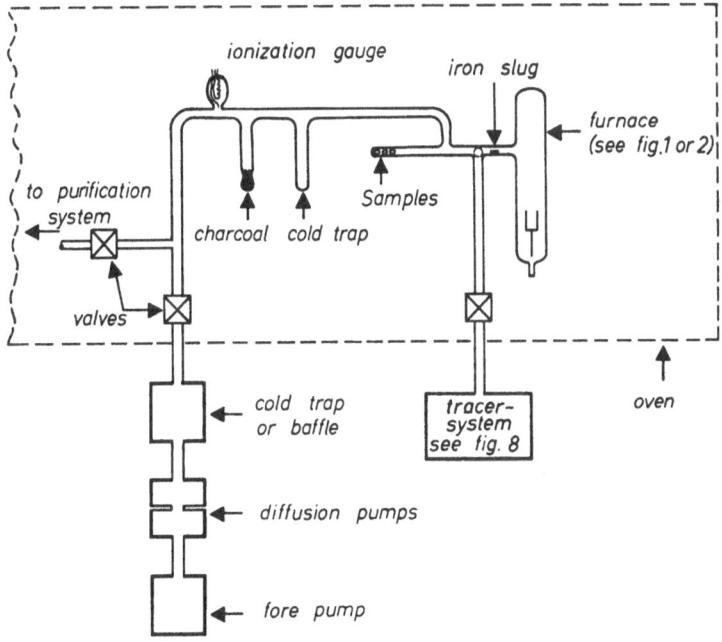

Fig. 3. Extraction system

from stresses. For this purposes several kinds of glasses are available: quartz, Supremax glass, Duran glass, Molybdenum glass, Kovar glass or Pyrex glass. In principle stainless steel or any similar metal could be used instead of glass. This is, however, more complicated to build and has no further advantage for argon measurement. Where a bakeout is desired, bakable metal valves with glass connections and a transition piece of Kovar glass are necessary. Otherwise greased joints or mercury valves are sufficient. For greased joints a high-vacuum grease with an extremely low vapor pressure (i.e., silicon or apiezon) must be used.

Mercury valves are able open or close a V-formed branched tube by raising or lowering the mercury level. The mercury has to be kept away

from the detection system and therefore a cold trap should be located at
the end of the extraction apparatus.

Fig. 3 shows a sketch of a heatable extraction apparatus.

For obtaining a vacuum a mechanical fore pump and two mercury
diffusion pumps with cold traps (which are necessary to prevent any
vapor of mercury in the extraction system) are used. They are cooled with
liquid air or liquid nitrogen. Oil diffusion pumps with a special aluminiumoxide baffle could also be employed. The connection between the pump
and the extraction system can be closed by a bakable metal valve. In the
vacuum system the pressure is measured by using an ionization gauge
or by a Pirany manometer. Normally, a pressure of about 10^{-6} mm Hg is
sufficient in the extraction apparatus.

It is advisable to run the diffusion pump continuously, but it is also
very important to clean the cooling baffle from time to time, especially
after having measured a sample very rich in gases.

A short rectangular glass sleeve is connected with the apparatus and
serves for storing the samples and preventing them from falling into the
crucible if a sudden air leak occurs. In order to avoid cutting and re-
sealing the glass apparatus every time a new sample is used, it is possible
to store ten to twenty samples in this manner, provided that the capacity
of the crucible is sufficient to melt them all in one run. Moving the sample
on the inside of the vacuum line is best performed magnetically with the
aid of a piece of iron. It may sometimes be useful to have a second storage
sleeve for changing the order of the samples.

During the extraction process, the escaping gases must be pumped
away continuously to avoid either discharges or a fast destruction of the
resistance heater. When using a high frequency-furnace, a purification
system consisting of a quartz pitcher containing zirconium or titanium
chips of 0.5 cm is placed just below the crucible. When these chips are
heated up either by lowering the high frequency-coil or by a second coil,
they act as getters to adsorb a large fraction of the active gases. Thus the
pressure in the apparatus is diminished to such an extent that no high
frequency discharges occur. The same effect can be obtained by the use
of a cold finger filled with charcoal and cooled with liquid air. The argon
and other gases will be adsorbed on charcoal and further high frequency
discharges will be prevented. Other methods of pumping away the escaping gases from the melting system consist of using either a small mercury diffusion pump (for non-bakable systems a Töpler pump is preferred)
or an additional purification furnace (described below). Zirconium and
possibly an additional furnace may lower the pressure sufficiently.

After the complete evacuation of the extraction apparatus, the sample
should be melted slowly to avoid the ejection of small particles. Normally,
complete degassing of the samples is obtained when the melting has

progressed about twenty minutes. Samples with a high amount of active gases are generally easier to degass than extremely gaspoor samples such as tektites. This is probably due to a rinsing effect.

C. The chemical flux method

Applying the flux method means that the samples are decomposed by chemical reactions with the aid of several suitable chemicals, the fluxes. Originally, this method was the only one used. Today, however, the hot extraction method is generally preferred. The main advantage of the flux method is that temperatures do not exceed 1000° C. These temperatures can be obtained by external heating. This means that the heating system does not necessarily have to be located within the vacuum system. Materials of lower melting point can be used. With a suitable flux, several minerals may react very slowly; sometimes it takes hours before a sample is completely degassed. Other minerals may react so fast that it is extremely difficult to hold such an exothermal process under control. Very often the material of the furnace or of the crucible will be attacked by the flux. A further disadvantage of this method is that due to the chemical reaction large amounts of gases, such as carbondioxide and sulphurdioxide are liberated and have to be absorbed by the purification system. Furthermore, every new mineral has to be tested to determine if degassing occurs quantitatively.

Incomplete degassing of the flux may be a serious drawback especially for the detection of very small amounts of gases. The flux method is nevertheless used.

The fused mass of high melting minerals or metals may sometimes not be sufficient for a quantitative degassing, tektites for instance must be vaporized for a complete extraction. In these cases the use of flux is recommended.

The following agents are frequently used as fluxes:

$NaOH$, Na_2O_2, $Na_2B_4O_7$, scarcely applied fluxes are:

Na_2CO_3, metallic Na, metallic Ca, $CaO/CaCl_2$ 1:1, $Na_2B_4O_7/Na_2CO_3$ 1:1.

For the dating of sylvite salt l. c. BUGGINGEN, SMITS and GENTNER (1950) dissolved the samples in water, which water was liberated from argon before this by bubbling carbondioxide through it. The argon was released from the water by the same process while boiling. It is also possible to use the fluxes in combination with one of the types of furnaces discussed previously. In this case the furnace can be used either with or without a molybdenum or a nickel crucible. When no crucible is used, a metal tube serves as furnace, but since the flux would attack any welded joints, the furnace must be machined from a single solid piece. For this purpose, a cylinder of stainless steel with a hole drilled trough it or a closed nickel tube are available and can be connected to the glass system via a cooled connection. A commercial furnace may be used to heat the cylinder.

This method is less sensitive, because the vacuum achieved is usually not very good. A rather large quantity of sample, up to fifty grams, should be taken. Due to the high volume of gas evolved from such a large sample a large dead volume should be connected with the apparatus in order to limit the pressure in the system. For this furnace the same construction could be used, as was the case for the heat extraction method.

When using the flux method, it is necessary to have a special device which allows the sample to be transfered into the crucible or the furnace within the vacuum system. The sample can be moved magnetically or it can be suspended above the crucible with a wire which can be cut by a current impulse. Such a device is necessary because the flux has to be melted and completely degassed for several hours with the pumps open. Fluxes having a very high vapor pressure in the molten stage (i.e. Na_2O_2) should be degassed just below the melting point or they will condense at the cooling flange. Next the flux will be melted under pressure and then cooled (see below). The sample can then be placed into the cold, clean flux and slowly heated. The ratio of flux to sample weights varies between $10:1$ and $1:5$. The proper quantity has to be determined for each type of sample. Time and temperature for the reaction are dependent on the conditions present in each case. Assuming an optimum combination of sample and flux, a time of two to three hours will generally be necessary for a temperature at about $100-200°$ C above the melting point of the flux. Whether an extraction is complete or not can be tested in the following way: The residual molten material is dissolved in water and examined for the presence of crystals, which were not destroyed during the heating.

D. Separation of argon from the extracted gas

The gases which have been treated by the first gettering in the extraction part have now to be separated from the other gases. This is done in the purification part of the apparatus which is connected to the extraction part by a valve.

First, the active gases will react with the chemicals located in the furnaces. Gases like water vapor or carbondioxide will be frozen out at a cold trap. The noble gases will be separated from each other by fractional adsorption by means of traps partly filled with adsorption material (charcoal). In principle, so-called molecular sieves could be used (WHIT-HAM, 1958). Depending on the detection method to be used this separation has to be carried out more or less quantitatively. Fig. 4 shows a purification apparatus which can be heated. There are no fundamental differences to non-bakable systems, the metal valves are merely replaced by greased joints or mercury valves. When using mercury, a cold trap is required at the end of the apparatus. The extraction system, the calibration system, the mass spectrometer and the diffusion pump are connected to the purification section by four valves. It is recommended that each part be separately connected to the pump. The furnaces are made of quartz or massive stainless steel or Kovar with a drilled hole, which contains the chemicals. Metal furnaces are connected with the

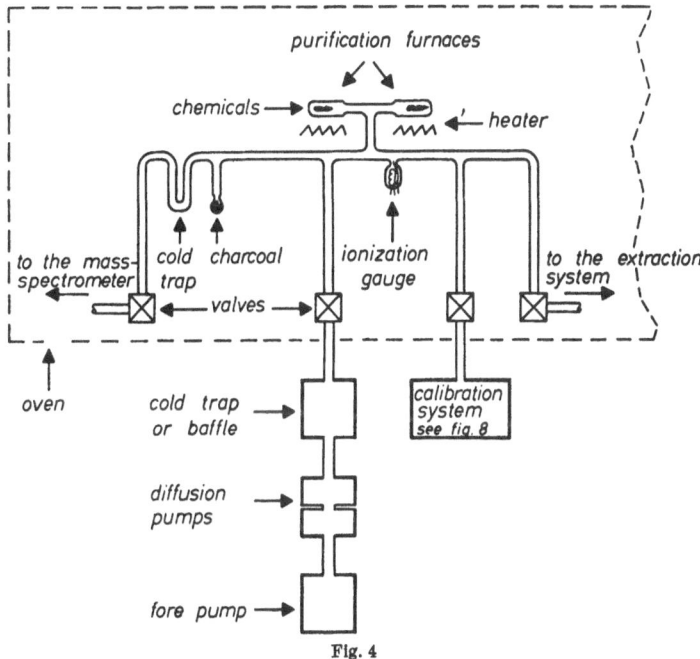

Fig. 4

apparatus horizontally by a metal-glass connection. For non-bakable apparatuses a cooled ground glass joint may be used. This arrangement is then mounted into a heater tube. The chemicals used are the following: Copper/copperoxide, calcium, titanium, less frequently barium, tungsten, uranium.

1. Copper/copperoxide furnace: A mixture of CuO powder and fine copper chips in the ratio 1 : 1 is used. At temperatures between 450° to 550° C hydrogen and oxygen react by forming water.

$$CuO + H_2 \longrightarrow Cu + H_2O$$
$$Cu + 1/2 O_2 \longrightarrow CuO$$
$$\overline{H_2 + 1/2 O_2 \longrightarrow H_2O}$$

After the furnace has cooled down this water can be frozen out in a cold trap. In addition CuO oxydizes CO to CO_2 which may also be frozen out. For a complete reaction, depending on the quantity of the gas, thirty minutes to three hours at reaction temperature will be sufficient. Before measuring, the furnace has to be cooled down in order to lower the pressure in the purification section.

2. Calcium furnace: At temperatures of $550-650°$ C, chips of calcium absorb most of the diatomic gases, mainly nitrogen and oxygen. The reaction time is about one hour. The bulk of the gases are absorbed quickly while the remaining gases, usually hydrocarbons, react slowly. Temperatures above $700°$ C should be avoided in order to prevent the decomposition of Ca_3N_2 and CaH_2. This procedure can be used at a later time to regenerate the consumed calcium. Before measuring, the calcium furnace should be cooled to $200°$ C in order to lower its vapor pressure to 10^{-8} mm Hg.

3. Titanium furnace: The titanium furnace is very universal because it reacts with practically all non-noble gases present, such as oxygen, nitrogen, carbondioxide, carbonmonoxide and hydrogen. Its disadvantage is the high reaction temperature. Usually a granulated titanium sponge acting as a getter for all non-noble gases except hydrogen, is used at $850°$ C for about an hour. After the reaction time the furnace will be cooled down to $350-450°$ C. At this temperature hydrogen reacts with the titanium. Further cooling to below $100°$ C causes the titanium vapor pressure to fall to 10^{-7} mm Hg.

4. Other types: Barium and calcium show the same chemical behaviour, with the exception that barium reacts more slowly. The advantage of this is the low vapor pressure. Metallic uranium is rarely used as getter material for temperature of $500-700°$ C. A tungsten filament directly heated to $2400°$ C is sometimes used for cracking hydrocarbon compounds.

The temperature of the furnaces may be regulated with a simple transformer and can be controlled by a thermocouple outside the furnace walls. To control the cleaning procedure, it is advisable to use an ionization gauge for pressures below 10^{-4} mm Hg. For higher pressures a Pirany gauge is very useful.

After the temperature of the furnace has dropped to room temperature, a U-tube is cooled with liquid nitrogen or liquid air ($-190°$ C). In about twenty minutes vapor pressure and residual hydrocarbons are frozen out. It is very important that the trap is free of any contaminants since even a tiny piece of dust is able to adsorb appreciably high amounts of argon. If the tube is clean no argon will be adsorbed. It may occasionally be expedient to use dry ice ($-79°$ C) for cooling.

The following analytical steps depend on the ultimate detection method to be used. If the absolute amount of argon has to be determined by mass spectrometry, it is not necessary to separate argon from the other noble gases. (Only for samples with a very high amount of helium, for example beryls, it is advisable to separate helium in advance in order to have a low pressure in the mass spectrometer tube.)

The argon is adsorbed on charcoal (which has been previously degassed at 400° C) when cooled with liquid air or liquid nitrogen for about fifty minutes. Within this time more than 99.9% of the total argon will be adsorbed. Then the valve which connects the extraction section with the purification section will be closed and the argon desorbes. The extraction section of the apparatus is thus disconnected from the spectrometer during measurement. This is recommendable because the pressure in the extraction section is usually one order of magnitude higher than that in the purification section.

For determining the argon volumetrically, it is absolutely necessary to separate the argon from the other noble gases quantitatively. For this purpose the apparatus has an additional valve or stopcock in order to close the charcoal finger from the other part of the apparatus. The argon will be adsorbed on this charcoal as described above and then the valve will be closed. Helium and neon will not be adsorbed, krypton and xenon, on the other hand, condense in the U-tube together with water. After this procedure the purification part will be evacuated completely. Besides argon the charcoal finger contains only a small amount of helium and neon, corresponding to the ratio of the volumes. After evacuation the pump will be closed and the valve to the still cold charcoal finger will be opened. By repeating this procedure the remaining helium in the volume of the cold finger can be reduced considerably. The amount of krypton which may also have been adsorbed is always negligible because the abundance of krypton in all rocks is considerably smaller than the abundance of argon.

E. Storage of argon samples

The extraction and purification systems may be connected with the detection system or may be completely separate. If the argon is determined by a mass spectrometer and the extraction and purification apparatus is directly connected to it, there will be long waiting times since the capacity of a spectrometer is more than ten times as large as the capacity of an extraction system. On the other hand it is possible to carry out argon separations using several independent extraction systems. The number of the analyses which can be done by a laboratory with only one mass spectrometer may thus be increased. For laboratories that do not have a mass spectrometer it is also possible to use a borrowed mass

spectrometer to measure their prepared samples. When applying the
volumetric method, it is preferable to combine the McLeod gauge
directly with the purification system. After having finished the determination of the absolute amount of argon with the McLeod gauge, it is
necessary to store the separated argon in order to determine the Ar^{40}/Ar^{36}-
ratio with a spectrometer. When using the isotopic dilution method the
McLeod gauge is not necessary.

In all these cases one needs a method to store the separated argon for
later measurement.

This can be done by small glass bottles in which the argon can be
collected and which can be closed off and then be separated from the appa-

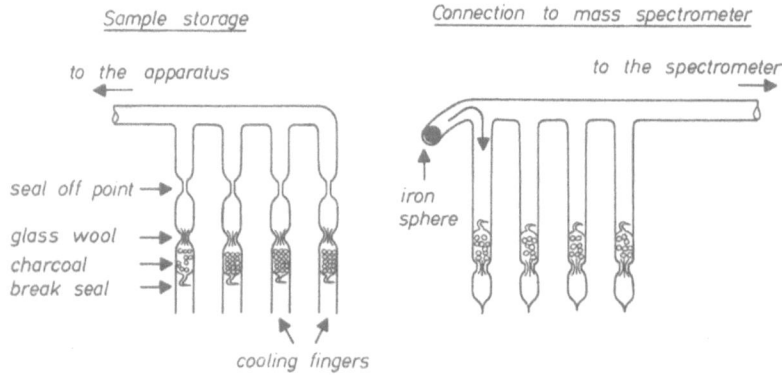

Fig. 5. Storage of argon samples

ratus. If the apparatus is bakable, the argon may be adsorbed by charcoal
in a separate finger which can be sealed off later. If the system is not
bakable, the collection can be done either by charcoal or by a mercury
Töpler pump and the separation from the apparatus may be accomplished
by closing a stopcock. Fig. 5 shows the principle of storage equipment for
bakable systems. The necessary number of cold fingers is attached to the
system and each finger is capable of being heated and sealed. The cold
fingers contain some grains of charcoal which are covered by glass wool.
At the center of the bottom is a friable capillary which is sealed off at
the end. The finger itself is prolonged at its lower end by a small open
glass tube. After the adsorption of the argon, the finger is sealed off
at the upper part while the argon is in the cooled charcoal. During
this procedure the vacuum will not be destroyed. The sample containers
are sealed and connected to the gas inlet system of the mass spectrometer. A steel sphere inside the tube moved magnetically can break
the capillary when dropped into the finger. Thus allowing the argon to
enter the mass spectrometer.

For non-bakable systems it is sufficient to use ground glass seals for a connection and a stopcock for closing and opening the system.

Additional literature to chapter II

1. Complete descriptions of the experimental procedure for argon extraction and separation: AMIRKANOV and BRANDT (1956), WETHERILL et al. (1956a), BAADSGAARD et al. (1957), CARR and KULP (1957), DAMON and KULP (1957a), LIPSON (1958), OKANO (1961), STARIK (1961), LONG and KULP (1962), BASSETT et al. (1963), KAWANO and UEDA (1964).

2. Partial descriptions of the experimental procedure and special details of argon extraction and separation: GOODMAN and EVANS (1941), ALDRICH and NIER (1948), GERLING and PAVLOVA (1951), HAYDEN and INGHRAM (1953), PANETH (1953), GENTNER et al. (1954a), AMIRKHANOV (1955), AMIRKHANOV et al. (1955a and 1955b), GENTNER and KLEY (1955), GENTNER and ZÄHRINGER (1955), WASSERBURG and HAYDEN (1955a and 1955b), ALDRICH (1956), ALDRICH et al. (1956b), DAMON and KULP (1957b), FIREMAN and SCHWARZER (1957), GENTNER and KLEY (1957), GENTNER and ZÄHRINGER (1957), SCHAEFFER and ZÄHRINGER (1959), SIGNER and NIER (1960), ZÄHRINGER (1960), HART (1961), HURLEY et al. (1961b), ZARTMANN et al. (1961), HINTENBERGER et al. (1962), McDOUGALL (1963c), GENTNER and LIPPOLT (1963), KIRSTEN et al. (1963), RANKAMA (1963), McDOUGALL (1964), MANUEL and KURODA (1964), ZARTMANN (1964).

3. Application of the flux method and further details: FARRAR and CADY (1949), GULDNER and BEACH (1950), MOUSUF (1952), GENTNER et al. (1953a), RUSSELL et al. (1953), WETHERILL (1953), GENTNER et al. (1954b), SHILLIBEER and RUSSELL (1954), SHILLIBEER et al. (1954), WASSERBURG and HAYDEN (1954a), KOLTZOVA (1955), THOMSON and MAYNE (1955), FOLINSBEE et al. (1956), LIPSON (1956), WASSERBURG et al. (1956), WETHERILL (1956), BAADSGAARD et al. (1957), REYNOLDS (1957), WASSERBURG et al. (1957), GEISS and HESS (1958), LIPSON (1958), HART (1961), DAMON et al. (1962).

4. Vacuum technique: ALPERT (1953), DIELS and JAECKEL (1962), DUSHMANN (1962), BARRINGTON (1963), GUTHRIE (1963), ROBERTS and VANDERSLICE (1963), TRENDELENBURG (1963), BECK (1964).

III. Determination of argon

A. Volumetric method

The volumetrical determination of the absolute amount of argon may be done precisely only with a McLeod gauge (ROSENBERG, 1938). This method is only applicable if the quantities to be measured are larger than 10^{-4} cm³ and therefore the sample weight has to be chosen accordingly. As pointed out previously, in most of the cases it is necessary to determine the Ar^{40}/Ar^{36} ratio with a mass spectrometer. For this purpose only a fraction of the gas is needed. Fig. 6 shows a common McLeod gauge. The McLeod volume, V_1, is connected with a selected capillary of ex-

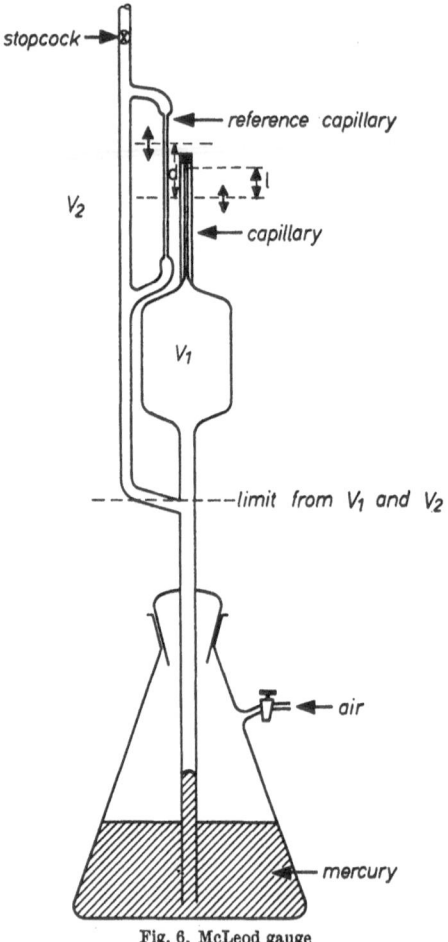

Fig. 6. McLeod gauge

tremely constant diameter $2r$. By moving a mercury level up and down the part $M_1 = M \dfrac{V_1}{V_1 + V_2}$ of the gas to be determined is compressed in the capillary. M_1 is the amount of gas present in the McLeod volume V_1, M the amount to be determined, V_2 the total volume of the apparatus except V_1. The difference, d, of the two mercury levels between the calibration capillary and the reference capillary, and the length l of the gas remaining in the calibration capillary (both in mm) has to be recorded.

M_1 can be calculated according to the relation $M_1 = k \cdot l \cdot d$, where k is a calibration factor for this McLeod which can be determined when V, p and T of a gas probe are known. ($V = \pi r^2 \cdot l$, p results from the equation $p \cdot V_1 = d \pi r^2 \cdot l$). Procedure for estimation from $\dfrac{V_1}{V_1 + V_2}$:

Measuring the pressure of a gas present in the McLeod before and after equilibrium with the total volume of the evacuated apparatus. A precise determination of the diameter of the capillary of the McLeod gauge is absolutely necessary because its radius influences the calibration factors. The best way to do it is to weigh a mercury column of a certain length. Very often separated small mercury spherules adhere to the inner wall of the capillary, this can be prevented by etching gently the inside of the capillary with dilute hydrofluoric acid in water. The rounded tip at the upper sealed-off end of the capillary is very difficult to calculate. The complication can be avoided when the capillary is cut off horizontally and closed with araldite. Preferably, the values d and l will be recorded at the moment when approximate $l = d$. When the pressure in V_2 is not negligible, eventually a correction will be necessary.

B. Mass spectrometric analysis

a) Introduction

In the following discussion, only the principle of mass spectrometric analysis will be treated. The treatment of technological details concerning the construction of spectrometers as well as the treatment of special types of spectrometers will not be given. There exist many textbooks where these subjects can be found: MAYNE (1952), EWALD and HINTENBERGER (1953), INGHRAM and HAYDEN (1954), DUCKWORTH (1958), HINTENBERGER (1962), BIRKENFELD et al. (1962), McDOWELL (1963), BRUNNÉE and VOSHAGE (1964). For routine analyses, one normally uses commercial spectrometers[1]. In the following it is assumed that the reader is familiar with the principles of mass spectrometry and only those technical details will be given which are concerned specifically with the argon- analysis in geochronology. The operation of a mass spectrometer and the specific difficulties in argon-analysis will be treated in greater detail.

b) Principle of the method

The principle is explained with the example of a single focusing Nier mass spectrometer (NIER, 1947) with an electron impact ion source, a 60° magnetic sector field and electron multiplier for ion detection (Fig. 7). These instruments have been very useful for age determinations.

---

[1] AEI, Manchester; CEC, Pasadena; CSF, Paris; Hitachi, Tokio; Leybold, Köln; MAT, Bremen; LKB, Stockholm; NAA, Pennsylvania; Thomson-Houston, Paris.

The purpose of the ion source is to ionize neutral gas atoms. Electrons are emitted from a tungsten cathode. With electric lenses the electrons are focused to a narrow beam and collected at a trap anode. Between the cathode and the anode they pass through a metal housing, the ionization chamber, where the gases are ionized. The production of ions is proportional to the partial pressure of the specific isotope. The ions have the potential of the ionization chamber. With a potential of a few volts the ions are extracted perpendicular to the electron beam. They then

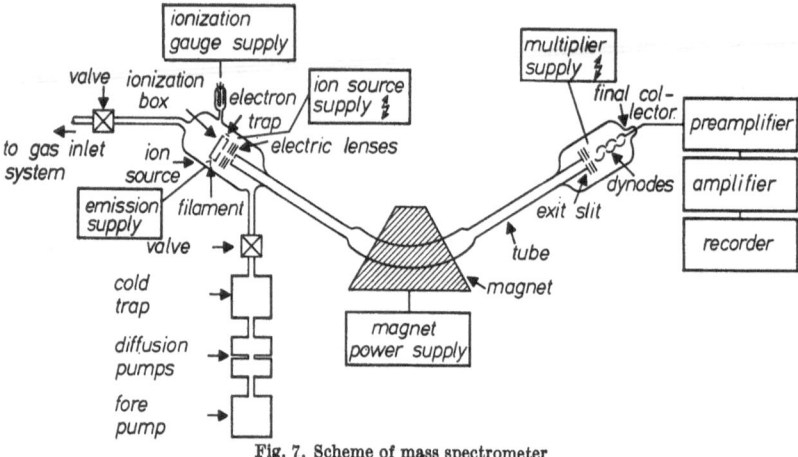

Fig. 7. Scheme of mass spectrometer

enter an electric lens system where they are focused to a narrow beam and accelerated to several kev. The acceleration voltage is much larger than the small potential differences due to the thermal energies of the atoms and their locations in the ionization chamber. The ion beam enters the mass spectrometer tube and passes through a magnetic sector field, the field lines of which are perpendicular to the ion beam. This sector field gives an optical picture of the entrance slit. The deflection of the magnetic field depends on the mass of the ion. The deflection radius r of the ions with mass m in the magnetic field B is given by $r = mv/eB$ where e is the charge unit and $v = \sqrt{2eU/m}$ the velocity (U is the acceleration voltage). The focal points with different masses can be moved along the exit slit by changing the magnetic field or the acceleration voltage. The collector current of a massline is a measure for the concentration of this isotope. The current is amplified and the signal then goes into a chart recorder.

c) Special details on argon analysis

The analysis of small amounts of rare gases is usually done with a magnetic 60° spectrometer. Sometimes the 180° spectrometer is used, which follows the same principles just described. The cycloid spectrometer and the omegatron have also been developed for age determination purposes. Cycloid spectrometer: The ions pass perpendicular to both a homogenic magnetic field and a homogenic electric field and follow cycloids, the focuses of which are mass dependent. This spectrometer has the advantage that its volume can be kept very small. The same is true for the omegatron (a time of flight spectrometer) in which the ions are accelerated in a homogenic magnetic field by high frequency voltage and where only those ions which are in the cyclotron resonance can reach the collector.

For normal conditions commercial spectrometers are fully satisfactory. Only for extreme requirements is a laboratory construction recommended (REYNOLDS, 1954 and 1956a; SMITS and ZÄHRINGER, 1955; SCHAEFFER, 1959).

The requirements for the vacuum of the spectrometer depend on the gas content of the sample which is to be determined. In most cases a final pressure of 10^{-8} mm Hg is sufficient. However, an absolute minimum would be a vacuum of 10^{-5} mm Hg. For this reason the whole spectrometer should be bakeable. A metal spectrometer tube with glass to metal seals suits this purpose very well. If no extreme strains are put on the vacuum technology, an all-metal spectrometer with flanges may be acceptable. The use of an ion getter pump is recommended in addition to the mercury diffusion pumps with a cold trap or oil diffusion pumps with a baffle. To prevent a pumping effect of the filament and a memory effect (see below), the emission current should not be above 0.5 mA. It should be possible to stabilize the emission current to 1 part in 10^4. The same requirements are valid for the acceleration voltage $(1-5$ kv). The focusing of the ion beam has to be adjusted from outside; the potentials of the lenses should consequently be variable. Lenses consisting of two separate parts which allow for independent adjustment of the potentials are recommended. For mass separation one should prefer an electric magnet to a permanent magnet, because large changes of the acceleration voltage cause mass discriminations (see below).

Common radii of the curvature are 25 cm, 6 and 4.5 inches. With a width of 0.3 mm for the entrance slit and 1 mm for the exit slit the mass resolution is about one hundred, which is good enough for argon determinations and which still gives optimal intensity. The use of a double collector system for Ar^{36} and Ar^{40} is advisable, especially when the spectrometer is operated under dynamic conditions (NIER, 1947; McKINNEY et al., 1950; SMITS and ZÄHRINGER, 1955). For the amplification of the ion current one uses either a vibrating reed electrometer or a d.c. electrometer. The amplifier should have 4 to 5 decades or a switch for changing input resistors to measure a large range of Ar^{40}/Ar^{36} ratios. An electron multiplier as preamplifier should only be used when very small argon contents are to be measured. As the Ar^{40}/Ar^{36} values are usually very large and since the sensitivity of a multiplier depends on the intensity, considerable correction has to be made. This is due to the mass discriminations arising from mass dependent electron emission at the first dynode and from the magnetic strewn field at the multiplier. A very commonly used multiplier is the system of commercial Cu-Be or Ag-Mg which is installed behind the exit slit within the vacuum system. The high voltage supply should be stabilized and designed for 2 to 3 kv and about 1 mA current. The amplified current is registered with a chart recorder. Recently electronic systems have been developed to transform the peak heights in digital numbers and to print them on punch cards. The evaluation can then be performed by computers.

The sensitivity is usually given in volts and can reach 10^{10} volt/cm³ argon. In practice, the smallest amount of argon to be detected is not given by the detection limit of the current but by the height of the background lines.

Mass discrimination: The ion current at the detector of a mass spectrometer is approximately proportional to the number of atoms of each isotope in the neutral gas. Therefore the measured current ratios give the true isotopic abundances. For more precise measurements some mass discriminating effects have to be considered.

A large mass discrimination is obtained by using a multiplier. The mechanism of electron emission on the first dynode appears to work in such a way that the isotope ions with constant velocity, v, produce nearly the same number of electrons. The ions, however, do not have constant velocities but constant energies $\frac{m}{2} v^2$. For this reason we have the equation $\frac{\varepsilon_1}{\varepsilon_2} = \frac{\sqrt{m_2}}{\sqrt{m_1}}$ for the specific sensitivity of detection, ε_i (volt/cm³), between two isotopes. For instance, Ar^{36} gives about a 5% higher current than Ar^{40} ($\varepsilon_{36} = 1,05 \times \varepsilon_{40}$). As there exist several other discriminating effects, it is necessary to determine empirically the sensitivity corrections by measuring a standard of exactly known isotopic composition. This has to be repeated after each re-adjustment of the spectrometer.

d) Operation of the mass spectrometer

α) Background lines, rest gas and its improvement

For a quantitative argon determination it is necessary that the isotopic abundances shown for the masses 36, 38 and 40 come exclusively from the argon of the sample, and that other types of ions with the same e/m-value do not make a contribution. Furthermore, argon not contained in the sample should be removed. In most cases there is a background for these masses which should be reduced so that the analysis is not affected.

Sources for background lines: The residual gas always contains a small quantity of chlorine and hydrogen. The chlorine and hydrogen will combine to form HCl, which will appear on the masses 36 $(HCl^{35})^+$ and 38 $(HCl^{37})^+$. Sometimes heavy organic molecules diffuse through the baffle. When using oil diffusion pumps lighter fragments are formed from these heavy molecules by the filament in cracking processes. Thus, we find, for instance, ions on mass 38 (f.i. $C_3H_2^+$).

When ionizing water vapor molecules, $(H_2^1O^{16})^{++}$ ions are formed. Some of them may be converted into $(H_2^1O^{16})^+$ ions by recombining after having been accelerated and before entering the magnetic field. The result of this will be that they appear on mass 36.

Another source of background lines is the so called memory effect. This is caused by some kind of exchange effect. Ions of previously measured samples are injected into slits on the ion source and on the collector plates. Under good vacuum conditions they do not appear in the spectrometer blank. When a sample is introduced, however, they are re-ejected and mixed with the new sample. This effect seems to be mass dependent and the ejection is most efficient with the same kind of gas. In the case of argon it is advisable to clean the spectrometer with atmospheric argon from time to time.

Another source is atmospheric argon from the residual vacuum. The contaminations of the extraction system give the same contribution since they are introduced with the sample. In the following, we will discuss these contaminations although they are not characteristic of the spectrometer. The Ar^{36} background presents a special difficulty for age determinations because an error in the abundance of this mass affects the determination of radiogenic argon by a factor of 296.

Improvement of the background: The background may be reduced by improving the vacuum and avoiding the use of chlorine containing agents while constructing the spectrometer and purification parts of the extraction system. Chlorine, hydrogen, argon and water vapor especially have to be reduced. The following procedures proved to be useful for reducing the background:

A) The whole spectrometer has to be baked up to 400° C for several hours for degassing the surfaces.

B) Degassing of the tungsten cathode and the vacuum tube filament over a longer period of time with higher emission currents and open pumps.

C) Continuous pumping for weeks and months.

D) Venting has to be restricted to an absolute minimum. If venting is absolutely necessary, it should be done with dry nitrogen or helium but *not* with air.

E) No use of ordinary oil diffusion pumps, only special oils such as silicon DC 705 with an aluminiumoxide baffle give satisfactory backgrounds.

F) No chlorine containing agents should be used when constructing and cleaning the spectrometer parts. It is of special importance that the spectrometer tube is not coated with tindichloride.

G) Occasional defrosting of the cold traps while the spectrometer is closed.

(These remarks are also valid for the extraction system.)

H) A memory effect can be diminished by running an intensive ion beam, especially with atmospheric argon.

I) Ar^{38} should not be used as a tracer if Ar^{38} of the samples is to be determined. (e. g. in the case of meteorites.)

K) Use of ion getter pump to decrease the hydrogen partial pressure.

L) Baking of the crucible and the zirconium getter over a long period of time and at higher temperatures than those used for the extraction.

M) Baking of the cleaning furnaces at higher temperatures. When using a calcium furnace, the formation of a calcium mirror has to be prevented, such a mirror getters hydrogen and, later on, releases it.

N) Baking of the cold traps and charcoal after each measurement.

O) No use of fluxes for the extraction.

P) If the apparatus is not bakeable use mercury-valves instead of greased stopcocks.

Q) Very careful cleaning of the gas.

R) Careful cooling of cold trap for condensing water and hydrocarbons.

S) During mass spectrometric analysis the titanium furnace may be kept at 300° C for continuous gettering of hydrogen.

T) For static measurement (see below) the spectrometer should be closed again immediately after complete gas inlet.

U) The samples themselves should be slightly baked (1 hr at 150° C). Inasmuch as this can be done without loss of argon and before this, they should, perhaps, be carefully etched (not with HCl!).

To judge the condition of the apparatus correctly, one performs the following measurements:

1. Background of the spectrometer (static measurement: with closed pump).

2. Background measurement with the evacuated extraction and purification systems.

3. Blank run, complete extraction and purification and mass spectrometric analysis without sample or with a low argon test sample.

The precision of the measurements depends on the result of such a blank run. If attention is paid to all points discussed, the following optimum results can be obtained with a specially adjusted spectrometer (the peaks were converted into the corresponding quantities of argon and are given in cm³ STP):

	Ar^{36}	Ar^{38}	Ar^{40}
Spectrometer blank	5×10^{-12}	2×10^{-12}	5×10^{-10}
Blank run	10^{-11}	10^{-11}	10^{-8}

In the evaluation the blank may be subtracted from the results obtained for the sample. The error, however, is rather serious because the runs with a sample are carried through under different conditions. Inserting the sample also involves a not quite definable change of the original background in the spectrometer.

β) Course of measurement

1. Gas inlet. Static method: The static method is used when only very small quantities of gas are at one's disposal (i.e., young samples or samples with small potassium concentrations).

The measurement is performed in a closed spectrometer. The spectrometer has to be absolutely tight and an ultrahigh vacuum must be achieved. The pressure in the ion source is the same as in the whole spectrometer. The pressure arising from the sample must not exceed 5×10^{-6} mm Hg. The gas inlet is just a pressure exchange between the purification section and the spectrometer. Even if the partial pressure is low, one minute is sufficient to reach equilibrium if the valve is opened properly. To introduce a relatively high fraction into the spectrometer, the volume of the purification system should be small. This technique has the advantage that there is no time limit for the measurements, if one disregards the pumping effect of the ion source.

Dynamic method: The dynamic method is only used for larger quantities of argon. The gas leaks in from a small storage volume with a pressure of about 10^{-1} to 10^{-3} mm Hg through a capillary or a leak in a membrane, thus maintaining for some time a rather constant pressure of 10^{-6} mm Hg while the spectrometer pump is open. A difference in pressure can also be maintained between the ion source and the spectrometer tube either by a tight formation or by a special exit slit (SMITS and ZÄHRINGER, 1955). When the isotope dilution method is used, the calibration follows from the isotope abundances. The calibration can also be done with a so-called "standard leak" which was calibrated with a known gas quantity at a known pressure in a reservoir. A double collector system for the masses 36, 38 and 40 is recommended for this method.

2. Analysis. For orientation in the spectrum, one can search for the intense background lines on mass 28 and 44. Usually, there are enough small peaks on the other masses for interpolation. Thus the magnetic current or the acceleration voltage can be marked with mass numbers. After gas inlet, the masses will be recorded at least ten times. This is usually done automatically. The magnetic current or the acceleration voltage has to be changed slowly and carefully so that the plateau of the peak is recorded clearly. The paper speed of the recorder has to be adjusted accordingly. Both sides of the peaks have to be recorded for determination of the zero level. If operated by hand the maximum value of a line and the zero level is recorded for several seconds alternatively. The range is selected in such a way that the height of the chart is optimally used. An amplifier with a range selector divided into decades and subdivisions is preferable. In modern commercial spectrometers the most favorable range is selected automatically. The range used and the mass number are marked on the chart for each peak. The following masses should be recorded:

$m = 36$: for determining Ar^{36} for the air correction.

$m = 38$: for additional control of the air correction or for calibration with Ar^{38} tracer.

$m = 40$: main peak for radiogenic Ar^{40}.

$m = 35$ and $m = 37$: for characterizing the chlorine background which allows one to estimate the HCl background lines. (The heights of the peaks are in proportion of $3:1$; this may help to orientate quickly in the spectrum.)

$m = 39$: The height of this line characterizes the purity of the gas of the sample (when Ar^{39} is not used as tracer).

Special attention has to be attached to the exact determination of the Ar^{40}/Ar^{36} ratio. These peaks have to be recorded repeatedly during the whole analysis.

First, the spectrometer background is recorded. The sample should be inserted while recording mass 36 to obtain the immediate difference between background and measured value. Ar^{36} has to be measured very carefully, because in most cases the accuracy of this measurement considerably influences the age determination.

After the isotope analysis, the calibration is performed unless the isotope dilution method is used (see below). An argon analysis is done in three steps:

1. Blank run. A complete run consisting of extraction, purification and isotope analysis has to be performed under the same conditions as if a sample is measured.

2. Main run. Same procedures as above with a sample.

3. Second run. In another complete run at somewhat higher melting temperature a test is made to determine whether or not the sample was completely degassed. The results for the blank run should be the same or nearly the same as the results for the second run.

3. Evaluation: When calculating mean values, attention has to be paid to possible shifts of the zero line. The mean values for background, sample and for the calibration gas are converted into gas quantities (see below). Before this can be done, the sensitivity of various isotopes has to be corrected, as already mentioned above. After determination of the gas quantities, the gas quantity of the blank run is subtracted. Usually, the gas quantities thus obtained are calculated for one gram of the sample material. Thus the Ar^{36} and Ar^{40} contents are expressed in cm^3 STP/g. The next step will be the correction of the atmospheric argon according to the equation:

$$Ar^{40}_{rad} = Ar^{40}_{total} - 295.6 \times Ar^{36} \ .$$

The same sample has to be measured several times if a correct age determination is to be obtained. Different sample weights should be used to be able to detect possible systematic errors. For estimating the error of an argon determination it is helpful to take into account the degree of agreement existing for several independent measurements of the same sample. Therefore, special care has to be taken to avoid systematic errors, especially argon losses during the extraction.

Presently some standard samples are in use which were analysed in different laboratories. They are of great help if the calibration is to be compared with other laboratories. Such standard measurements should be performed from time to time to assure that no serious systematic mistake is made (see below).

e) Appendix: Calculation of ages

The K-Ar age can be calculated from the Ar^{40} and K^{40}-content. The age is correct when the initial concentration of Ar^{40} was zero and when no diffusion losses took place.

The K-Ar age t can be calculated from the law of radioactive decay:

$$t = \frac{1}{\lambda} \ln \left\{ 1 + \frac{1+R}{R} \frac{(Ar^{40}_{rad})}{(K^{40})} \right\}$$

(Ar^{40}_{rad}) and (K^{40}) are given in number of atoms.

$\lambda = \lambda_{\beta-} + \lambda_k$ is the total decay constant from K^{40} and $R = \lambda_k/\lambda_{\beta-}$ is the branching ratio of the double decay of K^{40}.

For practical use the equation is changed as follows:

— Insertion of the most reliable values of the decay constants:

$$\lambda = 5{,}32 \times 10^{-10} \ y^{-1} \ , \quad R = 0{,}123 \ .$$

— Replacement of K^{40} through K_{total} using the isotopic abundance of K^{40} (0,0118%).

— Conversion of the ratio $(Ar^{40}_{rad})/(K)$ in $\dfrac{cm^3 STP}{g}\ \dfrac{Ar^{40}_{rad}}{K}$.

— Using common logarithms.

Thus, one arrives at the following equation:

$$t = 4320\ {}^{10}\log\left\{1 + 134,7\ \frac{Ar^{40}_{rad}}{K}\right\}$$

where t is given in $m.y.$ Due to the linearity of the function for small t, the formula $t = 2,53 \cdot 10^5 \cdot \dfrac{Ar^{40}_{rad}}{K}$ is a useful approximation, i. e. for ages up to 30 m.y. the error is less than one percent.

The relative error $\Delta t/t$ which arises from the relative errors from $\lambda = \lambda_{\beta-} + \lambda_k$ and $(Ar^{40}_{rad})/(K^{40})$, is given by the formula:

$$\frac{\Delta t}{t} = \frac{\Delta\lambda}{\lambda} + \frac{1-e^{-\lambda t}}{\lambda t}\ \frac{\Delta\{(Ar^{40}_{rad})/(K^{40})\}}{(Ar^{40}_{rad})/(K^{40})} + \frac{1-e^{-\lambda t}}{\lambda t} \cdot \frac{\lambda_{\beta-}}{\lambda}\left(\frac{\Delta\lambda_k}{\lambda_k} - \frac{\Delta\lambda_{\beta-}}{\lambda_{\beta-}}\right)$$

In practice the last turn can be neglected.

The error of the $(Ar^{40}_{rad})/(K^{40})$ determination enters with the factor $F = \dfrac{1-e^{-\lambda t}}{\lambda t}$, which is different for different age. With increasing age the influence of the analytical error decreases. The time dependence of F is as follows:

$$F = 1 - 2.66 \times 10^{-10} y^{-1} t + 4,72 \times 10^{-20}\ y^{-2} t^2 - 6.3 \times 10^{-30} y^{-3} t^3 + - \cdots$$

(Up to 100 m.y. is $F \sim 1$, by 1000 m.y. is $F = 0,78$, by 5000 m.y. is $F = 0,37$).

The error in the determination of the isotopic abundance of K^{40} and of the abundance of the argon-isotopes in the atmosphere can in comparison with the other errors be neglected.

Additional literature to chapter III

1. Volumetric method: PANETH and PETERS (1928), ROSENBERG (1939), GERLING and TITOW (1949), SMITS and GENTNER (1950), NODDACK and ZEITLER (1954), SHILLIBEER and RUSSELL (1954), HERZOG (1956), STARIK (1961).

2. Mass spectrometric analysis: ALLEN (1939a), ALLEN (1939b), THODE and GRAHAM (1947), ALDRICH and NIER (1948), SCHAEFFER (1950), SMITH (1951), INGHRAM et al. (1953), LIPSON and REYNOLDS (1954), REYNOLDS (1954), SCHAEFFER (1954), THOMSON and MAYNE (1955), AMIRKHANOV and BRANDT (1956), HARTMANN and BERNHARD (1957), GEISS and HESS (1958), NIER (1959), NIER (1960), ELLIOTT (1963), FARRAR et al. (1964), HINTENBERGER (1964a and 1964b), KAWANO and UEDA (1964).

IV. Calibration
A. General remarks

The absolute calibration of the noble gas isotopes is made by comparsion with known amounts of a standard gas sample. For a volumetric measurement the calibration is given by the dimensions of the McLeod gauge. It is advisable to check this mechanical calibration with several independent McLeod gauges. Two methods of calibration are applicable for mass spectrometric measurements:

(1) the peak height method (direct calibration), or (2) the isotopic dilution method. The choice between both depends on wether or not the same isotope being measured is used for the calibration. Both methods need small quantities of gas which must be exactly determined.

B. Preparation of gas standards with an exactly determined amount of gas

A glass apparatus with the following components is used:
— Reservoir containing the calibration gas.
— McLeod gauge for the determination of pressure.
— Titanium furnace to purify the gas.
— Charcoal trap for the argon adsorption.
— Pumping system consisting of fore-pump and diffusion-pumps.

A frequently used reservoir is a glass bulb with a sealing off point at the one end and a break off tube at the other end. The volumes of those bulbs are determined by filling them with quantities of mercury which are weighed afterwards (approximately 100 to 200 cm³). Several of these bulbs are then attached to the system by fusing. After complete evacuation of the apparatus, the calibration gas is added until a pressure of about 0.1 to 1 mm Hg is reached. The gas will be cleaned from contaminations by using a titanium furnace at 850° C for several hours. After the furnace has cooled to room temperature, the argon will be adsorbed on charcoal to check if it is free from helium and neon. The residual pressure in the apparatus must be negligible. When the argon is clean, it will be desorbed from the charcoal and the pressure in the apparatus should be measured with the McLeod gauge at least twenty times. In addition, the temperature should be measured in order to convert the results to standard conditions. When the calibration bulbs are sealed off, great care has to be taken that the temperature in the bulb does not increase. It is best to place the bulbs in a horizontal position.

Volume dilution

First, the standard method used by GENTNER and ZÄHRINGER (1957) will be described (Fig. 8). The standard sample S is connected to a five liter bulb B of known volume. A U-shaped glass capillary C (the volume of which is determined with mercury (approximately 0.1 cm³) is then attached. This capillary can be closed at both ends by using the glass tubes T and the ground glass joint. The glass tube is filled with mercury to the point that it closes the capillary in its vertical position. The capillary depression of mercury prevents the latter from entering the capillary.

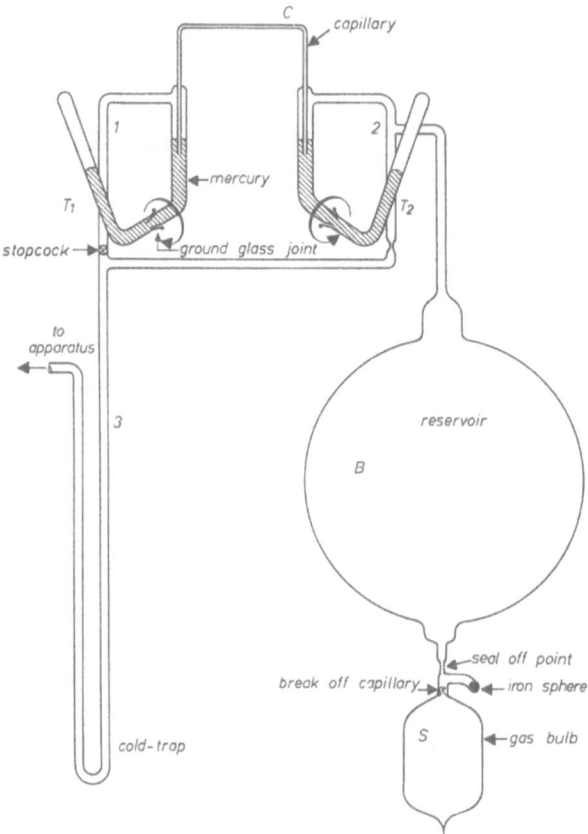

Fig. 8. Volume dilution system

This kind of closing the capillaries is preferred because argon could be dissolved in the grease on the ground glass joints. The supply *1* joins supply *2* via a seal-off point that emerges from the other end of the capillary. Disconnection of supply *2* is possible. The main supply *3* that results now forms a U-tube which serves as a trap to freeze out the mercury and is connected to the purification part of the apparatus via a valve.

The system is quickly evacuated; the pipette closed, and the reservoir filled with the standard argon. The standard bulbs can be sealed off after opening. In this way the gas pressure is diluted by a factor of fifty. This procedure is necessary, because it is very difficult to reliably determine such a low pressure with a McLeod gauge. A further dilution is made with the pipette. The ratio of the volumes $V_{capillary}$ to $(V_{bulb} + V_{connections})$ determine the capillary gas content. When the dimensions of the system just described were used, the gas content is about 10^{-6} cm^3 STP. If a pipette is taken off, the U-tube is cooled with liquid air. Then T_2 will be opened and a half minute's waiting is necessary to reach the equilibrium pressure. T_2 will then be closed and T_1 is opened; now the calibration gas diffuses via purification system into the mass spectrometer. The joint lead to the system should be as small as possible, so that the volume of the whole apparatus is not changed during the calibration. Unfortunately the mercury has to be pumped off from the U-tube and its vapor pressure spoils the vacuum in the purification system. This disadvantage is evaded by the most recent development that consists in using a metal pipette instead of a glass pipette. For this purposes, a hole of about one mm^2 cross section is drilled trough a massive stainless steel piece of ten cm length. This small hole can be closed on each side by silver seals attached to membranes. When calibrating, care must be taken to notice how far the silver seals arch into the capillaries. This may greatly influence the accuracy of the calibration.

This dilution method has a great advantage: it is possible to take out hundreds of standards without changing the amount of gas present significantly. For a V_{bulb} to $V_{capillary}$ ratio equal to 50000 the decrease in the argon content is about one per cent after using fivehundred standards.

The calibration technique by WASSERBURG and HAYDEN (1955a) is more elaborate. A series of calibrated bulbs with break seals are connected to a reservoir containing a known amount of gas. After breaking the seal of the reservoir, it is necessary to wait until pressure equilibrium is attained. All the break off tubes are then simultaneously closed by raising a mercury level. The separation from the system is done by melting the glass connection. These bulbs are now directly connected to the purification system. For each calibration measurement one of these bulbs has to be opened.

Precision

The following values determine the calibration errors:

— Pressure determination.

— The calibration of the applied McLeod gauge.

— The determination of the various volumes.

— Purity of the calibration gases.

Careful work allows one to reduce the calibration error to less than one per cent.

It is absolutely necessary to employ different McLeod gauges and different balances for the calibration of the McLeod capillaries and the other volumes.

All interested laboratories can check their calibration by using suitable standards available at special offices (i.e. Nat. Bur. of Stand., Washington, Mineralog.-Petrogr. Institut, Bern, see JAEGER et al. (1963) or US Geological Survey, see LANPHERE and DALRYMPLE (1965)).

C. Direct calibration with "pipettes"

This calibration process is performed by pipetting Ar^{40} directly into the spectrometer.

Consider the Ar^{40} which has already been measured. Its peak height has the value m_1, which increases to m_2 after introducing the calibration argon. From this the absolute amount of the gas of the sample (M in cm^3 STP) is calculated: $M = V \dfrac{m_1}{m_2 - m_1}$. V represents the volume of calibration gas reduced to standard conditions. This method is very exact since the argon from the sample and the calibration argon are both measured at the same pressure.

Sometimes the calibration gas is introduced after the gas extracted from the sample is pumped away. In this case the sensitivity of the spectrometer has changed because of the varied pressure conditions. The most accurate calibration can be made by using a calibration volume between 50 and 200 per cent of the volume delivered by the sample.

Usually, no Ar^{40}, but argon with isotopic composition of air is used for this calibration method.

Due to the low amount of Ar^{36} in atmospheric argon, a direct calibration of the Ar^{36} will not be exact when the Ar^{36} portion of the calibration gas is used. It is preferable to calibrate Ar^{36} by using the sensitivity ε_{40} obtained from the calibration of Ar^{40}. In this case it is necessary to consider the sensitivity correction factor $\varepsilon_{40}/\varepsilon_{36}$. The measurement of the ratio Ar^{40}/Ar^{36} in the pipette is used to check the constancy of the sensitivity ratio. This ratio has to be known because of the necessary air correction. When only the amount of Ar^{40} is to be determined, the direct method is recommended. In these case no correction of the sensitivity has to be made since all calibrations are done with the same isotope. If more precision is required, consecutive measuring of several pipettes will increase the accuracy and simultaneously check the linearity of the spectrometer.

D. Isotope dilution method

An absolute amount of an isotope is determined by adding a known amount of another isotope of the same gas. The known isotope serves as a tracer (RITTENBERG, 1942; INGHRAM et al., 1950; INGHRAM, 1953;

INGHRAM, 1954; AMIRKANOV et al., 1955b). This procedure reduces the problem to one of determinating an isotopic ratio. The determination of an isotopic ratio is much more precise than the calibration method described above. Another advantage of this method: systematic errors (i.e., gas losses) are the same for both gases. A calibration using this method cannot be influenced by systematic errors if the tracer is added properly to the sample gas during the extraction procedure. For this purpose the calibration section will be connected directly to the extraction system.

Isotopic dilution requires only a small amount of tracer isotope in the sample. As a tracer pure Ar^{38} is especially suitable because it has the smallest abundance in the sample. The Ar^{36} of the sample is the most convenient for correcting the atmospheric component. If Ar^{38} is added in quantities comparable to the Ar^{40}, the initial Ar^{38} content can be neglected. The amount of radiogenic Ar^{40} in the sample can be calculated:

$$[Ar_{sample}^{40}] = [Ar_{tracer}^{38}] \left(\frac{Ar^{40}}{Ar^{38}}\right)_{mixture} \cdot \frac{\varepsilon_{38}}{\varepsilon_{40}}$$

where ε_{38} and ε_{40} represent the specific sensitivities for argon isotopes (see chapt. III) and $\left(\frac{Ar^{40}}{Ar^{38}}\right)$ the measured peak height ratio. The square brackets give the amount of argon in the tracer and the sample.

In samples with a high atmospheric argon content a small correction depending on the Ar^{36} content has to be applied to the Ar^{38} peak. For this case the radiogenic argon content can be calculated from the following formula where the symbols have the same meaning as above (296 and 5.35 are the abundance-ratios of Ar^{40}/Ar^{36} and Ar^{36}/Ar^{38} respectively in atmospheric argon):

$$[Ar_{rad.}^{40}] = [Ar_{tracer}^{38}] \left\{ \frac{\frac{1}{296} - \left(\frac{Ar^{36}}{Ar^{40}}\right)_{mixture}}{\left(\frac{Ar^{38}}{Ar^{40}}\right)_{mixture} \left(\frac{1}{296} - \frac{1}{5.35 \times 296} \left(\frac{Ar^{36}}{Ar^{38}}\right)_{mixture}\right)} \right\}.$$

In a similar way, LIPSON (1958) has calculated a formula for the case in which the tracer is a mixture of all three isotopes. Ar^{38} is usually used as a tracer. This isotope can be produced by thermocolumns and bought at the Helvetian Institute of Technology at Zurich. Ar^{38} can also be made from a chlorine-containing compound by $Cl^{37}(n, \gamma) \, Cl^{38} \xrightarrow[37 \, min.]{\beta^-} Ar^{38}$ reactions. Smaller amounts are produced in a mass separator.

Potassium containing glass can be irradiated with fast neutrons in a reactor. Ar^{39} with a half life of 260 y is homogeneously produced by the $K^{39}(n, p) \, Ar^{39}$ reaction in the glass. Its Ar^{39} content can be determined in a mass spectrometer. An ideal tracer in powdered form is obtained in this way and can be added in small amounts to the samples (NAUGHTON, 1963).

Additional literature to chapter IV

1. Direct calibration: SCHAEFFER (1959), SCHAEFFER and ZÄHRINGER (1960), MERRIHUE et al. (1962), HEYMANN et al. (1966).

2. Isotopic dilution method: HAYDEN and INGHRAM (1953), BA-LESTRINI (1954), WASSERBURG and HAYDEN (1954a), WASSERBURG and HAYDEN (1954b), CARR and KULP (1955), WASSERBURG and HAYDEN (1955a), FOLINSBEE et al. (1956), WASSERBURG and HAYDEN (1956), WETHERILL et al. (1956a), CARR and KULP (1957), DAMON and KULP (1957a), DAMON and KULP (1957b), REYNOLDS and LIPSON (1957), WEBSTER (1959), GERLING (1961), OKANO (1961), STARIK (1961), DAMON (1962), LONG and KULP (1962), WASSERBURG et al. (1962), BASSETT et al. (1963), WANLESS et al. (1965).

V. Neutron activation analysis

In special cases the K-Ar age determination may be carried out by means of the neutron activation technique (CURRAN, 1953; MOLJK et al., 1955; STOENNER and ZÄHRINGER, 1958; ARMSTRONG, 1959; WÄNKE and KÖNIG, 1959).

The main advantages of this method are:

1. Extremely high sensitivity.

2. Contaminations after the radiation do not influence the result.

3. Smaller amount of equipment as compared with the mass spectro-metric method.

4. It is possible to carry out simultaneous measurements of argon and potassium on the same sample.

The experiments must be carried out near a reactor because of the short half life of Ar^{41} (1.82 hr). The method is based on the n-capture of Ar^{40} and Ar^{36}. The decay of the radioactive isotopes Ar^{41} and Ar^{37}, produced by n-capture is used to determine the Ar^{40} and Ar^{36} content in the sample.

The following reactions are important for the K-Ar age determination:

Main reactions

$$(1) \qquad Ar^{40} (n, \gamma) Ar^{41} \xrightarrow[\beta^- (1.25\,\text{Mev})]{1.82\,hr} K^{41} \text{ (with slow neutrons)},$$

$$(2) \qquad Ar^{36} (n, \gamma) Ar^{37} \xrightarrow[e^- \text{-capture}]{35\,d} Cl^{37} \text{ (with slow neutrons)},$$

$$(3) \qquad K^{39} (n, p) Ar^{39} \xrightarrow[\beta^- (0.57\,\text{Mev})]{260\,y} K^{39} \text{ (with fast neutrons)}.$$

Side reactions

$$(4) \qquad Ca^{40} (n, \alpha) Ar^{37} \xrightarrow[e^- \text{-capture}]{35\,d} Cl^{37} \text{ (with fast and slow neutrons)},$$

(5) K^{41} (n, p) Ar^{41} $\xrightarrow[\beta^{-}(1.25\,Mev)]{1.82\,hr}$ K^{41} (with fast neutrons) ,

(6) Ca^{44} (n, α) Ar^{41} $\xrightarrow[\beta^{-}(1.25\,Mev)]{1.82\,hr}$ K^{41} (non effective) .

The way in which the activation is performed depends mainly on the calcium content of the sample. Calcium poor samples such as chondrites and iron meteorites can be irradiated before the gas is extracted. The argon and potassium is determined in the same specimen. The potassium can be detected in two different ways: (1) by the reaction $K^{41}(n, \gamma)$ K^{42}, (2) by reaction 3. In the latter case the powdered sample is packed in a cadmium foil and placed in a small evacuated quartz ampoule. The sample is then exposed to two irradiations:

(1) exposure for several weeks to built up Ar^{39} activity with fast neutrons. The irradiation is disconnected and Ar^{41} decays.

(2) exposure to slow neutrons (without cadmium foils) to produce Ar^{41} and Ar^{37} from Ar^{40} and Ar^{36} respectively.

After the irradiation, the gas from the sample will be extracted, cleaned and measured.

The Ar^{37} decay is a K-capture with low energy. Best counting efficiency is obtained when the argon is mixed with the filling gas of the counter. For Ar^{41} and Ar^{39} an another technique has been used for dating stone meteorites (by WÄNKE and KÖNIG, 1959). The argon is filled in a small box with a thin mylar window. This container is placed directly before an end window counter. An aluminium absorber of $14\,mg/cm^2$ can be used which is able to absorb the Ar^{37} activity without influencing the other activities. The separation of the activities of Ar^{39} and Ar^{41} is easy since they have very different half-lifes. The absolute calibration is performed by irradiating simultaneously argon and potassium standards.

In samples with unfavorable calcium content the argon has to be extracted before irradiation (ARMSTRONG, 1963). It is filled in small quartz containers and only the gas is exposed to neutrons. Special care has to be taken that the container is free of calcium and that it gives no contribution to the Ar^{37}.

With a neutron flux of 10^{12} to 10^{13} neutrons/cm^2 sec an activity of 1 to 5 imp/min corresponds to an amount of 10^{-8} cm^3 Ar^{40}, three hours after the end of the irradiation. Generally irradiation times of several days are used to built up the Ar^{37} activity.

Additional literature to chapter V

JENKINS and SMALES (1956), HERR (1960), WÄNKE (1960), KRANKOWSKY and ZÄHRINGER (this book, last chapter).

Potassium Analysis

By

O. MÜLLER

I. The physical and chemical properties of potassium and its isotopes

Potassium belongs together with lithium, sodium, rubidium, cesium, and francium to Group I A of the Periodic System. All the alkali elements are univalent, electropositive, and form strong bases. The alkali salts and hydroxides are in general easily soluble in water and are almost completely dissociated in aqueous solution. Indeed, lithium differs somewhat from the other alkalies resembling more the alkaline earth elements. In nature, sodium and potassium are the most abundant of the alkalies.

The alkalies are the most closely related group of elements in the Periodic System. Therefore, the analytical chemistry of the alkali elements always poses some problems. Quantitative procedures often involve complicated separation steps for the determination of single alkali elements.

Some physical properties of the alkali metals are summarized in Table 1.

Table 1. *Physical properties of the alkali metals* (from KOLTHOFF and ELVING, 1961)

	Lithium	Sodium	Potassium	Rubidium	Cesium
Atomic number	3	11	19	37	55
Atomic weight	6.940	22.991	39.096	85.48	132.91
Melting point, ° C	179	97.8	63.7	39.0	28.5
Boiling point, ° C.	1336	880	760	700	670
Hardness	0.6	0.4	0.5	0.3	0.2
Density, g./cc.	0.530	0.963	0.857	1.894	1.992
Heat capacity, cal./g.	0.941	0.293	0.18	0.08	0.0482
Atomic radius, A.	1.55	1.90	2.35	2.48	2.67
Radius of cation, A.	0.60	0.95	1.33	1.48	1.69
Oxidation potential, v. . . .	3.06	2.72	2.93	2.99	3.04
Ionization potential of gaseous atom, v.	5.37	5.12	4.32	4.16	3.87
Volume of ion, 10^{-23} ml . . .	0.14	0.37	0.99	1.36	1.95
Thermal conductivity of liquid metal, cl./cm/sec	0.5919 (183° C)	0.2055 (100° C)	0.1073 (200° C)	0.07 (m.p.)	0.044 (m.p.)
Electrical resistivity of liquid metals, μohms	45.25 (230° C)	9.65 (100° C)	13.16 (64° C)	23.15 (50° C)	36.6 (30° C)
Electrical conductivity (Cu = 100)	18	37	24	14	8
Latent heat of fusion, cal./g. .	158	27.05	14.6	6.1	3.77
Latent heat of vap., cal./g.. .	4680	1005	496	212	146
Volume change on fusion, vol. %	1.5	2.5	2.41	2.5	2.6

For absolute dating work the knowledge of the isotopic composition of an element is of great importance. Table 2 shows the stable and radioactive nuclides of potassium presently known. Also the short-lived nuclides of potassium are included. They are of importance for the neutron activation technique.

Potassium has three naturally occurring isotopes with the mass numbers 39, 40, and 41, the abundances of which are 93.08, 0.0119, and 6.91% respectively (NIER, 1950). K^{40} is unstable and decays by K-electron capture to Ar^{40} and by β^--emission to Ca^{40}. The determination of the branching ratio of the K^{40}-decay is difficult; it is discussed by HOUTERMANS in this book.

Table 2. *The stable and radioactive nuclides of potassium*

Nuclide	Half-Life	Mode of Decay	Energy of Radiation MeV	Produced by
K^{37}	1.3 s	β^+	β^+ 4.6	K-γ-2n
K^{38}	7.7 m	β^+	β^+ 2.8 γ 2.16	Cl-α, n K-n-2n K-p-pn K-γ-n Ca-d-α
K^{39}	stable			
K^{40}	1.29×10^9 y	β^- K, γ	β^- 1.33 γ 1.46	natural source end-product Ca^{40}, Ar^{40}
K^{41}	stable			
K^{42}	12.4 h	β^-	β^- 3.58, 1.99 γ 1.51	Ar-α-pn K-d-p K-n-γ Ca-n-p Sc-n-α
K^{43}	22.4 h	β^-	β^- 0.81 etc. γ 0.63, 0.4 etc.	Ar-α-p
K^{44}	17.0 m	β^-	β^- 4.9, 1.5 γ 1.13 etc.	Ca-n-p
K^{45}	20.0 m	β^-	β^- 2.1, etc. γ 0.176, 1.7 etc.	Ca^{48} (d, nα) Ca^{48} (α, α2np)
K^{47}	17.5 s	β^-	β^- 4.1 etc. γ 0.6, 2.0, 2.6	Ca^{48} (γ, p)

The radionuclide data have been compiled from STROMINGER et al. (1958) and "Nuclear Data Sheets" (1958—1960), K^{45} from MORINAGA and WOLZAK (1964), K^{47} from KUROYANAGI et al. (1964).

II. Terrestrial and extraterrestrial abundance of potassium. The potassium minerals

Potassium as a lithophile element is considerably enriched in the upper earth mantle comprising about 2.6% of the lithosphere. Since soils and sedimentary clays absorb potassium salts more strongly than sodium salts, the potassium content of sea water is substantially lower (0.038%). The average Na/K ratio is 1.09 in magmatic rocks, 27.8 in sea water, and 0.36 in sedimentary clays (HILLER, 1962).

Potassium occurs, apart from the salt minerals, mostly in natural silicates and is a major constituent of feldspars and micas.

The feldspar family comprises of the following minerals:

Orthoclase	$K[AlSi_3O_8]$	monocline
Microcline	$K[AlSi_3O_8]$	tricline
Albite	$Na[AlSi_3O_8]$	tricline
Anorthite	$Ca[Al_2Si_2O_8]$	tricline
Leucite	$K[AlSi_2O_6]$	tetragonal
Nepheline	$KNa_3[AlSiO_4]_4$	hexagonal

Very pure orthoclase with a crystal habitus, which differs from usual feldspars, is called adular. Sanidine is a rapidly crystallized potassium feldspar not separating sodium and a disordered Al-Si distribution. The Ca-Na feldspars, anorthite and albite, are completely soluble in each other. The intermediate compounds are called plagioclase.

The mica group comprises mainly the following potassium minerals:

Muscovite	$KAl_2[(OH, F)_2AlSi_3O_{10}]$	monocline
Phlogopite	$KMg_3[(OH, F)_2AlSi_3O_{10}]$	monocline
Biotite	$K(Mg, Fe, Mn)_3[(OH, F)_2AlSi_3O_{10}]$	monocline

The lithium containing minerals, lepidolite and zinnwaldite, also belong to this group.

We can obtain information on the extraterrestrial abundance of elements by studying meteoritic matter. Since the chondrites are the most common representatives of the various meteorite types, their composition should best approximate the relative abundance of elements in our solar system.

The average potassium content of chondrites is assumed by various authors as follows:

GOLDSCHMIDT (1937)	UREY (1952)	LEVIN et al. (1956)	MASON (1962)	
ppm K				atoms per 10^6 atoms Si
1540	900	900	1000	4300

The value given by GOLDSCHMIDT is too high, while those of the other authors are in rather good agreement. EDWARDS (1955) and EDWARDS and UREY (1955) use a high temperature distillation process for isolating potassium and flame photometry for measuring it. The result of this procedure is a uniform potassium distribution in common chondrites, averaging 880 ppm K for 32 falls, with a range from 750 to 1050 ppm. This range was confirmed by KIRSTEN et al. (1963) using isotope dilution technique, and by WÄNKE (1961) applying the neutron activation method.

The achondrites, however, are notably deficient in potassium, and the above distribution is less uniform. The range is from 400 ppm to 10 ppm in the diogenite Johnstown.

The carbonaceous chondrites show figures ranging from 100 to 1400 ppm K (MASON, 1963 b).

The amphoterite-chondrites are a subclass of the olivine-hypersthene chondrites. They are characterized by iron-rich olivine and orthopyroxene, and a low content of nickel-iron. Chemical analyses were reviewed for 12 amphoterites by MASON and WIIK (1964). Potassium (and argon) determinations of 18 amphoterites have been elaborated by KAISER and ZÄHRINGER (1965) using isotope dilution technique. Large variations of potassium were found: from 200 ppm (Ensisheim) to 2180 ppm (Soko Banja), both within single as well as for different amphoterite specimens. Potassium argon dating work on iron meteorites by neutron activation yielded extremely low potassium contents of the metal phase ranging from several ppb to several hundred ppb K. In schreibersite, troilite, and silicate inclusions, however, potassium is enriched to several ppm, STOENNER and ZÄHRINGER (1958), WÄNKE (1961), MÜLLER and ZÄHRINGER (1966).

Detailed information of the abundance of elements in meteoritic and stellar matter is available by ALLER (1961), MASON (1962a), SUESS and UREY (1956), UREY and CRAIG (1953), UREY (1952).

III. Chemical methods of potassium separation and determination

1. Precipitation reactions and gravimetric techniques

The classical chemical method for quantitative analyses of potassium is the precipitation and weighing of an insoluble potassium compound of stoechiometric composition. The best results are obtained with potassium perchlorate $KClO_4$, potassium hexachloroplatinate K_2PtCl_6, potassium tetraphenylboron $(C_6H_5)_4BK$, and potassium silvercobaltnitrite K_2Ag $[Co(NO_2)_6]$. In all cases mainly rubidium, cesium, and ammonium do interfere, the precipitates of Rb and Cs are even heavier soluble than those of potassium. Conversely, the natural occurrence of Rb and Cs is considerably lower than that of K. The interference of Rb and Cs can therefore be neglected in potassium analyses for most rocks, minerals, tektites, and meteorites, used in potassium argon dating.

Perchlorate method

Potassium perchlorate is an unhygroscopic compound having thermal stability up to about 350° C. The solubility in water is rather high: 1.68 g are dissolved in 100 g of water at 20° C. The solubility decreases rapidly when absolute ethanol is the solvent: 0.012 g of $KClO_4$ are dissolved in 100 g at 25° C. This value is lowered to 0.0045 g in n-butanol. In waterfree ethylacetate $KClO_4$ is practically insoluble (WILLARD and SMITH, 1922, 1923). These authors also studied the solubilities of the other alkali and earth alkaline-perchlorates in organic solvents. Their results showed much higher solubilities than that of $KClO_4$. Aluminium and iron perchlorates are also soluble in absolute ethanol. The determination of potassium as perchlorate is therefore possible in the presence of Li, Na, alkaline earths, Al, Fe. Sulfates and ammonium salts must be absent if the perchlorate method is used. The sulfates can be removed with barium chloride, because barium perchlorate is soluble in alcohols.

The potassium perchlorate may also contain rubidium and cesium perchlorates. These latter elements can be determined by a different procedure (O'LEARY and PAPISH, 1934).

Chloroplatinate method

This method is based on the insolubility of potassium chloroplatinate in ethanol. The corresponding lithium and sodium salts are soluble. However, rubidium, cesium, and ammonium also form insoluble chloroplatinates. Since the composition of the K_2PtCl_6 is stoechiometric, there are three ways to determine the potassium:

a) weighing of the K_2PtCl_6,
b) reduction of the salt to platinum and determining it,
c) determining the KCl content.

The solubility of potassium chloroplatinate is comparatively low in ethanol (Table 3), ARCHIBALD et al. (1908).

Table 3. *Solubility of K_2PtCl_6 in ethanol at 20° C*

ethanol weight-%	mg K_2PtCl_6 in 100 g ethanol
80	8.5
90	2.5
100	0.9

The influence of anions is as follows: nitrates, bromides, and iodides should be converted to chlorides by repeated evaporation with hydrochloric acid. Sulfates interfere in the presence of sodium, because sodium sulfate is insoluble in ethanol. The alkaline earths, with the exception of barium, form ethanol soluble chloroplatinates; therefore, there is no interference in the presence of magnesium, calcium, and strontium.

Tetraphenylboron method

The sodium tetraphenylboron is an ideal precipitating reagent for determining potassium, rubidium, and cesium. The advantages of this method are as follows: The potassium salt corresponds stoechiometricly to the formula $(C_6H_5)_4BK$. Its solubility is extremely small compared with other slightly soluble potassium salts. This compound is therefore also very appropriate for determining trace amounts of potassium. Its thermal stability is good. The high conversion factor of the precipitate to potassium is extremely favorable. Most cations and anions do not influence the precipitation of potassium. The solubility of potassium tetraphenylboron is 0.58 mg in 100 ml of water at $20°$ C. This value is even lower if the reagent concentration of the solution is about 0.1 % (WITTIG and RAFF, 1950; RAFF and BROTZ, 1951; GEILMANN and GEBAUHR, 1953).

Cobaltnitrite method

Potassium is quantitatively precipitated in an acetic or weak nitric acid solution with a sodium cobaltnitrite reagent. A disadvantage of this method is that the composition of the precipitate is not well defined. It varies between $K_2Na[Co(NO_2)_6] \times nH_2O$ and $K_3Na_3[Co(NO_2)_6]_2 \times nH_2O$ depending on temperature, concentration of reagent and amount of the alkali present. It is necessary to maintain consistent working conditions and to establish empirical factors in this gravimetric procedure.

When silver cobaltnitrite (BURGESS and KAMM, 1912) is used as a precipitation reagent, a remarkably consistent composition of the precipitate is achieved. The application of $K_2Ag[Co(NO_2)_6]$ also increases the sensitivity for potassium.

(Sources for III.1.: KOLTHOFF and ELVING, 1961; SCOTT's Standard Methods of Chemical Analysis, 1962; TOLLERT, 1962.)

2. Ion exchange chromatography

In section III.1. it was explained that the separation of alkali metals with precipitation reactions is difficult. Ion exchange chromatography enables the analyst to separate chemically related elements more easily (for example, the separation of rare earths).

Synthetic organic ion exchange resins are usually used for analytical separation (ADAMS and HOLMES, 1935). These resins consist of a three dimensional polymere network which carries the active groups. Acid groups ($-SO_3H$, $-COOH$, $-PO_3H_2$, phenolic $-OH$) make a cation exchange possible by replacing hydrogen. Basic groups cause an anion exchange in the strongly basic quarternary ammonium group, or an acid absorption if the group is only slightly basic (amino-group). Fig. 1 shows the way of synthesis and the chemical structure of a sulfonic acid resin and a strongly basic ion exchange resin.

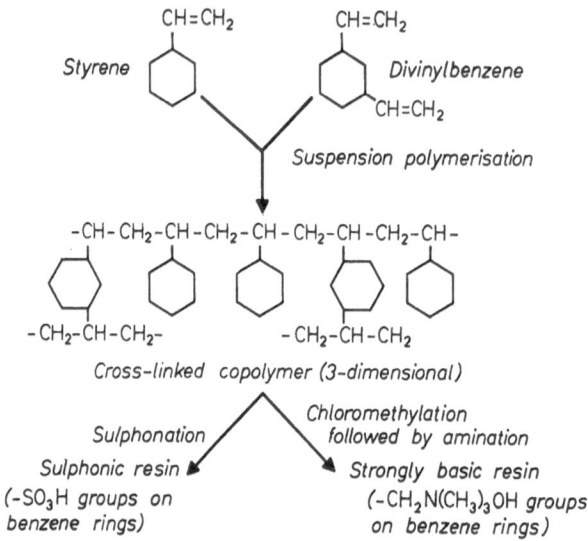

Fig. 1. Preparation of synthetic ion exchange resins (from CORNISH, 1960)

The resins are prepared by polymerizing styrene and divinylbenzene (DVB). The proportion of DVB can be varied considerably (usually 8—12%). The resins of this type are characterized by their DVB content. Most of the inorganic reagents cannot cause the resins to change. Application of concentrated acids, however, cause strongly basic resins to change slightly. In their OH^--form they are not indefinitely preservable when exposed to oxygen and sunlight. Some of the reliable types of synthetic resins are listed in Table 4.

Table 4. *Some commercially available ion exchange resins*

Manufacturer	Sulphonated cation exchangers	Strongly basic anion exchangers
Dow Chemical Co.	Dowex 50	Dowex 1
Rohm u. Haas Ltd.	Amberlite IR-120	Amberlite IRA-400
Chemical Process Co. . . .	Duolite C-20	Duolite A-42
Farbenfabriken Bayer . . .	Lewatit S-100	Lewatit M 500
Permutit Co. Ltd.	Zeocarb 225	Deacidite FF
Farbenfabrik Wolfen . . .	Wofatit KPS-200	

These products are on the market in different grain sizes (specifications usually in mesh) with varying contents of divinylbenzene. The acid resins are in the H^+-form, the basic resins in the Cl^--form [for instance, Dowex 50, 50—100 mesh, X-8, H^+-form (X-8 = 8% DVB)].

Inorganic as well as organic ion exchange resins are used for analytical separations (i.e. synthetic zeolite, aluminum oxide, ammonium-12-molybdophosphate, zirconium phosphate, and zirconium tungstate).

To carry through separations of different ions, the ion exchange equilibrium has to be known. Almost all ion exchange reactions are reversible and the equilibrium is determined by the properties and relative quantities of the components present in the system. The exchange of ions takes place strictly stoechiometricly.

The equation

$$A + B_r \rightleftharpoons A_r + B$$

may stand for the exchange of two univalent ions where A and B are the ions exchanged. Index r refers to the resin phase, absence of index is indicative of the ion in the solution surrounding the resin. The selectivity coefficient, K, describing the position of the equilibrium, is a measure for the affinity of an ion to the exchanger resin, and is defined for univalent ions by the equation:

$$K_B^A = \frac{[A]_r \times [B]}{[B]_r \times [A]} \ .$$

The equation for exchange reactions of ions with different charges is:

$$K_B^A = \frac{[A]_r^b \times [B]^a}{[B]_r^a \times [A]^b} \ .$$

Parantheses are concentrations: in the solution (without index), and in the resin phase (index r). Lower case a and b are absolute values for the charge on A and B respectively.

The selectivity coefficient is calculated from experimental data. It is important to know it for evaluating ion exchange processes. $K_B^A > 1$ means that a larger portion of A is absorbed by the resin than of B, $K_B^A = 1$ says that there is no selectivity for both ions.

Detailed theoretical papers on ion exchange equilibrium and applications on ion exchange separations have been published: EKEDAHL et al. (1950), GREGOR (1951), HÖGFELDT (1952), BAUMANN and ARGERSINGER (1956), RICE and HARRIS (1956), SCHINDEWOLF (1957), SAMUELSON (1963).

There are two techniques which may be used when working with ion exchangers: the batch method and the column method.

Batch operation

If the batch method is used, the resin and the solution are in one vessel. Both are mixed by stirring until equilibrium is reached. The solution is filtered off, and the resin prepared for the next cycle. The operation is then repeated. The extent to which the ion exchange takes place depends on the resin selectivity of the specific ions. This procedure utilizes only part of the resin's total capacity.

Column operation

Much better results can be obtained, by using this method. Fig. 2 shows two standard types of ion exchange columns.

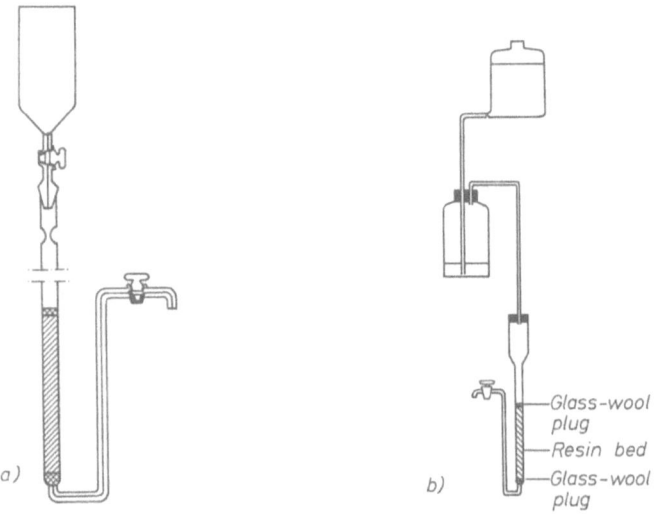

Fig. 2. a) Column used particularly for routine analyses. b) Column of standard type with applied pressure

An ion exchange column may be compared to the theoretical plates of a distillation column and requires analogous mathematical treatment. Since the exact mathematical calculation of the plates for the ion exchange column is very complicated, it is seldom used.

The process of the mutual separation of exchangeable ions by an ion exchange column is called chromatography. The resin bed serves as a chromatographical medium. The solution containing the ions to be separated is poured on the resin. The separation may then take place in the three following ways:

a) Displacement

The following experiment may exemplify this: A solution of sodium chloride is poured on a cation exchange column in the H^+-form. A calcium chloride eluant then flows steadily through the bed. This causes the sodium ions to move downward in a sharply bound band. In the eluant appear hydrochloric acid, sodium chloride, and calcium chloride (in that order) if the selectivities increase in the same order.

b) Elution

The reverse of a displacement is called elution. In this case the ion form of the resin is the same as the ion in the eluant. The eluant always has a smaller selectivity than the ions to be separated. A mixture of calcium chloride and sodium chloride solution may, for instance, be separated by elution with hydrochloric acid (resin in the H^+-form). Sodium chloride is eluted first. Elution of calcium chloride follows. Both salt solutions are mixed with hydrochloric acid. This technique is very often used for analytical separations.

c) Frontal analysis

The steady flow of an ion solution through a resin bed is called frontal analysis. This process may include either displacement or elution, or both. Ions appear in the eluant in the same order as their increasing selectivities. This method is usually used for removing an undesired component from the analysis solution and for desalting tap water.

Separations of alkali elements with cation exchangers

The separation of alkali elements from one another is generally based upon elution from strongly acid cation exchangers (for example, Dowex 50, with dilute hydrochloric acid). The selectivity of the resin increases with

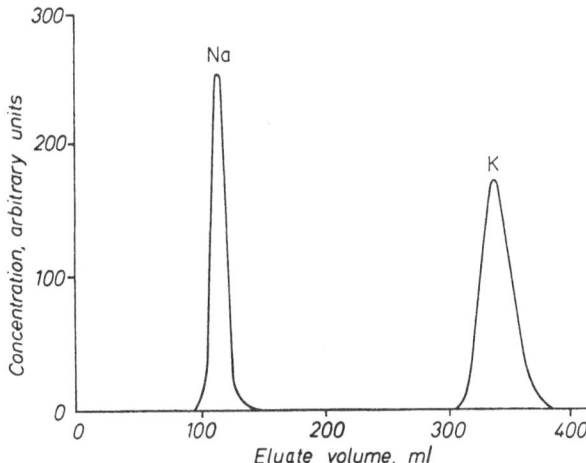

Fig. 3. Separation of 3 mg of sodium from 9 mg of potassium by elution with 0.6 n hydrochloric acid at room temperature. Column: diameter 9 mm, length 200 mm, Resin: Dowex 50, 200—400 mesh, X-12, H^+-form, Flow rate: 0.97 ml cm^{-2} min^{-1}

atomic weight. The alkali metals are eluted as separate bands in the order: Li, Na, K, Rb, Cs. Fig. 3 shows the separation of sodium from potassium with Dowex 50 (CORNISH, 1958).

Sometimes stepwise or gradient elution with increasing eluant concentration is used (COHN and KOHN, 1948; SWEET et al., 1952). A separation of Li, Na, and K can easily be performed within 8 hrs (SWEET et al., 1952).

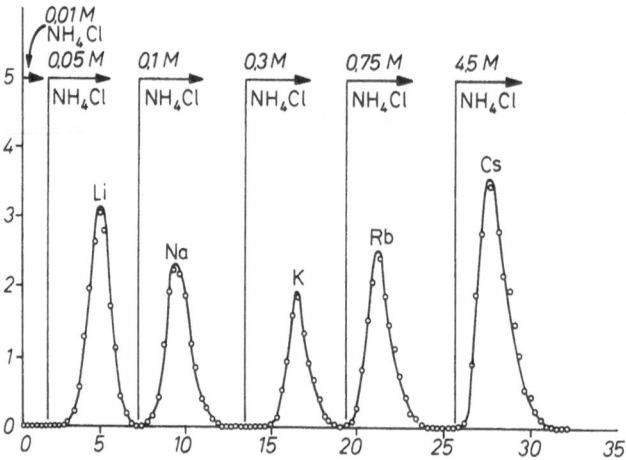

Fig. 4. Separation of the alkali metals. (Eluate concentration versus eluate volume.) Zirconium tungstate column: diameter 4 mm, length 123 mm, Flow rate: 0.75 ml cm^{-2} min$^-$

In addition to organic resins inorganic ion exchangers (such as zirconium phosphate and zirconium tungstate) can be used for the mutual separation of alkali metals. Fig. 4 illustrates the results of KRAUS et al. (1956).

In many applications, for example in mineral analyses for age determinations, not only a mutual separation of alkali metals is desired, but also a separation from other elements, such as calcium, strontium, aluminium, and iron which are held more strongly by a cation exchanger in a hydrochloric acid medium, ALDRICH et al. (1953). Complete separation is essential to the mass spectrometric analysis of rubidium and strontium. Because the abundant Ca^{40} would interfere with K^{40} in the mass spectrometer, the separation of potassium and calcium is necessary in an isotope analysis of potassium. The recovery of K, Rb, Sr, Ba, and rare earth elements from stony meteorites for mass spectrometric measurements was described by SCHUMACHER (1956b) using the cation exchanger Dowex 50.

Separations of alkali elements with anion exchangers

The alkali elements are unaffected by anion exchangers in most media. Thus anion exchangers are used in many group separations of alkali metals from other elements. Of some interest is the procedure published by SAMUELSON et al. (1955): They separated and determined the alkali metals in the presence of Ca^{2+}, Mg^{2+}, V^{4+}, Fe^{3+}, Al^{3+}, Cu^{2+}, Ni^{2+}, Co^{2+}, Mn^{2+}, and Zn^{2+} by using a Dowex 2 anion exchanger in the EDTA-, acetate-, and oxalate-form. This method may be of interest in potassium analysis of samples with low potassium content where the other elements would interfere with the quantitative detection of potassium. In radio-chemical procedures (neutron activation) this ion exchange technique can be used to separate the alkali elements from other contaminants.

3. Flame photometry

The prime advantages of flame photometric methods for the determination of alkali elements are summarized as follows:

1. Contrary to gravimetric techniques it is possible to determine alkali (and nonalkali elements) in the same solution of the sample without any chemical separation, or at least, with only one rough precipitation of iron, aluminum, calcium, etc.

2. As long as the solution of the standard is of similar composition, the flame photometric determination of the alkali elements can be carried out in the presence of a great number of other cations and anions.

3. Because of the high sensitivity, the amount of sample required for flame photometry is smaller than that used in gravimetric methods. This advantage is decisive in dating work of meteorites, tektites, and rare terrestrial material, where only a limited amount of sample is available.

4. Flame photometry is time saving especially when a larger number of alkali determinations are required.

5. The precision obtainable in flame photometry equals and sometimes exceeds that of other chemical methods. This is particularly true when low concentrations of alkali metals are present.

6. The ability to differentiate between potassium, rubidium, and cesium is a remarkable feature of flame photometry.

The determination of alkali elements by flame photometry is not absolute and requires relative comparison to standard solutions. The composition of these standards must always be the same as that of the sample solution. To adjust the standard to the sample solution is difficult when samples of unknown composition must be analyzed. Thus in some cases it would be better first to remove disturbing elements by precipitation with ammonia and/or ammonium carbonate.

The choice of optical system depends on the composition of the sample to be analyzed and the sensitivity required. Two types for the flame analysis are used: filter photometers and monochromators. Filter photometers are more sensitive. However, they are incapable of separating the radiation from two elements whose spectral lines are close together. For example, the rubidium line at 780.0 mμ can effect the potassium line at 768.2 mμ. If separation of these lines cannot be achieved by a filter photometer a monochromator must be used and it is further necessary to operate with narrow slits.

There are two ways of measuring the alkali elements:

1. Direct measurement of the potassium intensity.

2. Measurement of the intensity ratio between the potassium line and a supply line. This is the so-called Internal Standard Method. The advantages of this method are: mutual interference of various elements in the flame is largely reduced. Errors due to variations in the viscosity and surface tension of sample and standard solution are greatly eliminated.

Opponents maintain that

1. the internal standard cannot prevent mutual radiation interference of certain elements,

2. the addition of another element increases the background, and

3. errors may result if the sample unexpectedly contains the internal standard as an impurity. No fundamental reasons exist to give one of the both methods the preference.

The consideration of cation and anion interferences, mutual line interference, interference by mutual excitation or suppression, and interferences due to band spectra is very important in flame photometry. It is not possible to compare data reported in the literature, because the type of filter photometer (or spectrometer), the burners and fuels used have a large influence on the results. For example, the presence of magnesium in potassium solutions causes depression of potassium emission in the Beckman photometer with hydrogen-oxygen flame but no such effect is observed with the Perkin Elmer propane-air flame (COOPER, 1963).

Also the presence of anions can considerably influence the potassium emission: free acids, phosphate, carbonate, bicarbonate, borate, and oxalate ions decrease the potassium radiation.

For further information on flame photometric analysis some bibliographies are cited:

BURRIEL-MARTI and RAMIREZ-MUÑOZ (1957), DEAN (1960), HERRMANN (1956), HERRMANN and ALKEMADE (1960), KOLTHOFF and ELVING (1961), LUNDEGARDH (1929, 1934), SCHUHKNECHT (1961).

IV. Physical methods of potassium determination
1. X-ray fluorescence analysis

Each element has a characteristic X-ray spectrum, which can be generated by bombardment with high-speed electrons or high-energy photons. Fluorescent X-ray spectrography is concerned with analyzing the characteristic spectrum of a sample. It is termed "fluorescent", because the method of exciting the sample is by primary X-rays. This method is non-destructive, sensitive and rapid, and the physical state is immaterial. In contrast to optical radiation, X-radiation involves high-energies and is generally limited to wavelengths between 0.3 and 5 Å.

The X-rays are usually produced by X-ray tubes. High speed electrons hit a metal target producing an X-ray spectrum. This spectrum consists of a continuous background and characteristic wavelengths of the target material.

Two basic features of a sample's fluorescent radiation make a chemical analysis possible: first, the wavelength is indicative of which elements are in the sample, and secondly, the intensity of each line is proportional to the percentage composition of that element. The basic arrangement of a spectrograph's components is schematically represented in Fig. 5.

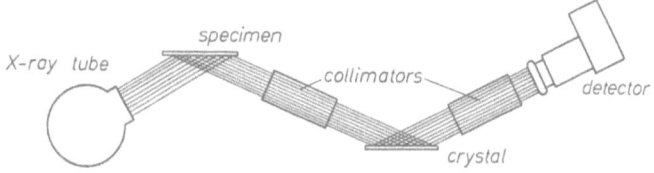

Fig. 5. Principle of an X-ray fluorescence spectrometer

X-rays interact with the specimen to be analyzed. This incident radiation excites the specimen's characteristic fluorescent radiation which is emitted in all directions. This radiation must be collimated into a parallel beam. The beam then strikes the crystal which will diffract only one wavelength for each setting (BRAGG's law). To measure the whole spectrum, the crystal is turned through an angle of 90 degrees. The detector turns at twice the crystal speed so as to intercept the diffracted radiation. The entire analyzing system is evacuated to eliminate air absorption of the long wavelength fluorescent radiation. This is important for the analysis of the light elements with an atomic number < 22. The lightest element, which can be analyzed with a normal X-ray fluorescence equipment, is magnesium.

Selection of the tube's target material is important in fluorescent excitation for several reasons. First, it imposes its characteristic spectrum on the tube's continuous spectrum. This is significant since the most effective wavelength for exciting an element is about 0.2 Å shorter than the element's absorption edge. However, the limited choice of X-ray tube targets requires emphasis to be placed on the tube's broad continuous spectrum instead of several characteristic wavelengths. Because the continuous spectrum's intensity increases with atomic number, it is wise to choose the target material with the highest possible atomic number. Tungsten is usually chosen as target material, since it not only has a high atomic number but also a high melting point permitting greater tube current. However, for potassium and all light elements, a chromium tube is more effective than that of tungsten.

The analyzing crystals are an important part of the spectrograph, since both the intensity and resolution are controlled by the crystal's properties. When a parallel beam strikes the crystal face, only certain wavelengths will be diffracted according to Bragg's law:

$$n \lambda = 2d \sin \theta$$

where n is the order of diffraction, λ the wavelength, d the crystal's interplanar spacing, and θ the angle between the crystal surface and the incident radiation. Bragg's law indicates that wavelengths greater than $2d$ cannot be diffracted by a particular crystal. Intensities and mechanical considerations usually limit the maximum wavelength to less than $1.8\,d$. The angular separation of two wavelengths increases as d decreases. Thus, for maximum resolving power, a crystal with an interplanar spacing as small as possible should be chosen. Generally, the alkali halides, such as LiF ($2d = 4.03$ Å), are used. LiF, however, is useful only up to a maximum wavelength of about 3.7 Å: For lighter elements with longer wavelengths, crystals such as ethylenediamine-di-tartrate (EDDT) ($2d = 8.81$ Å) or pentaerythrite (PET) ($2d = 8.74$ Å) are in use. For extremely long wavelengths, such as that of magnesium's K line, a Gips crystal with a large interplanar spacing ($2d = 15.18$ Å) is necessary.

In addition to plane crystals, curved crystals can be used to "focus" radiation diverging from a point or a linear source. The primary advantage of curved crystals: only a very small amount of specimen is necessary to give intensities comparable to much larger sample quantities in a plane crystal spectrometer.

The three types of detectors most commonly used in X-ray spectrometers are the Geiger, the proportional and the scintillation counter.

By using pulse height analyzing equipment, a window can be set such that only those amplitudes between the voltage V_1 and V_2 will be recorded. Thus when one wants to analyze potassium, most of the calcium peak can be eliminated by discrimination. The statistics of pulses are such, however, that no amount of discrimination alone can separate the peaks of directly neighboring elements. Elements separated by two or three atomic numbers can be well resolved if their concentrations are about equal and more than a few percent. When the concentration of an element is below 1 % it is difficult to distinguish from the tails of neighboring major peaks.

The specimen preparation is an important precondition for exact chemical analysis. Since 99 % of the incident X-rays penetrate the specimen only to a depth of about 50 μ, the quantitative results depend mainly on the surface layers. Great care must, therefore, be taken to make sure the surface of the specimen is representative of the specimen as a whole. Although samples can be prepared in many ways, not all insure homogeneity and few exhibit a linear relation between counting

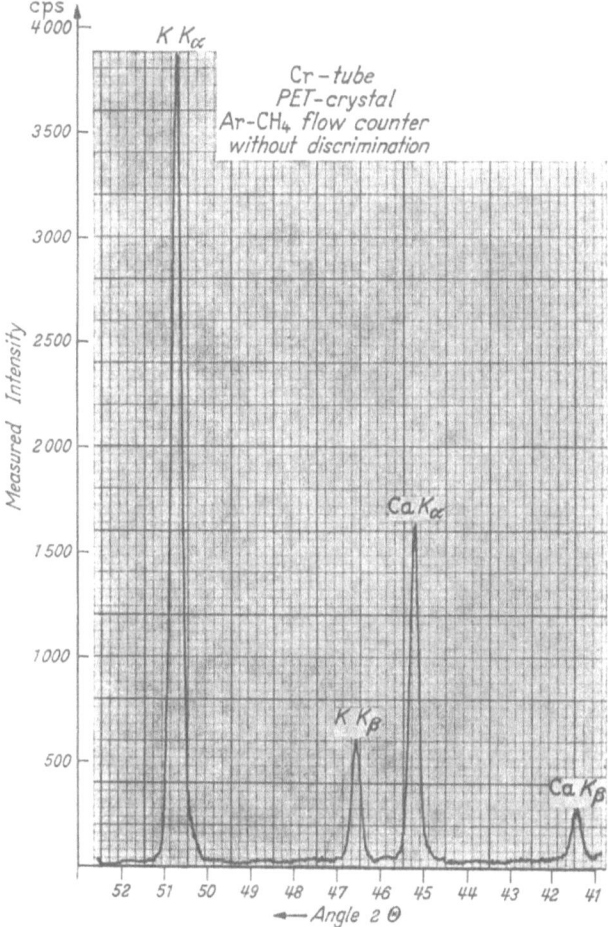

Fig. 6. X-ray fluorescence spectrum of potassium and calcium in granite G-1 (4.58% K; 0.99% Ca)

rate and percent composition. The following method of specimen preparation for rocks, minerals, tektites, and other glasses developed by ROSE et al. (1963), offers a homogeneous sample whose constituents show a linear relation to counting rate for silicate materials of different chemical composition.

The specimen to be prepared is mixed as a fine powder with La_2O_3 and $Li_2B_4O_7$ in a ratio of $1:1:8$, respectively. The La_2O_3, which is a heavy absorber of X-rays, minimizes differences in absorption between samples of different composition. When the mixture is heated sufficiently, the $Li_2B_4O_7$ acts as a flux to fuse the mixture, thus eliminating the hetero-

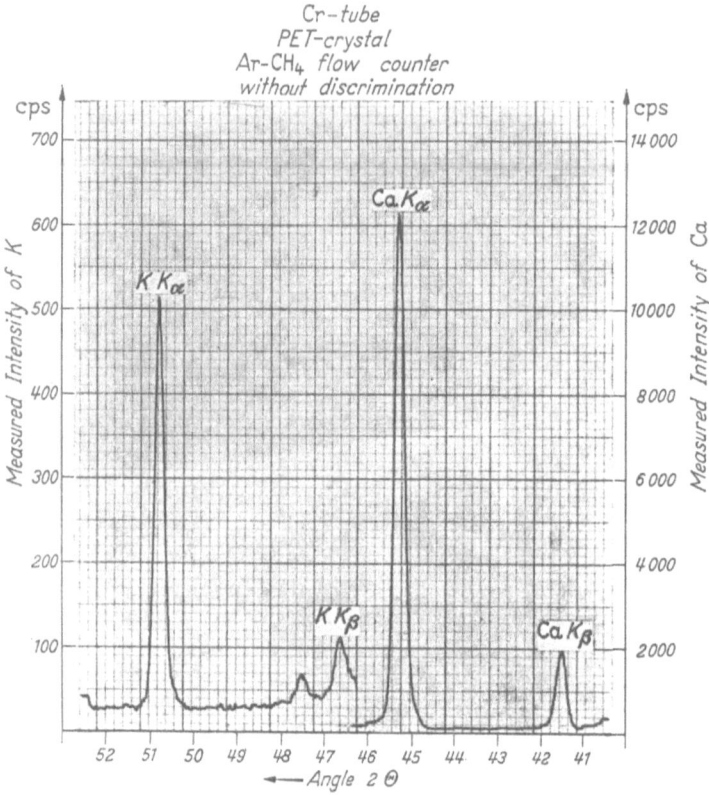

Fig. 7. X-ray fluorescence spectrum of potassium and calcium in diabase W-1 (0.54% K; 7.83% Ca)

geneity of composition between grains and within a grain itself. The mixture is heated in a graphite crucible first to about 750° C to remove any CO_2 and H_2O present. The temperature is then raised to 1100° C for about 10 minutes. The fused material is allowed to cool in the crucible. The bead thus formed still exhibits some inhomogeneities. Therefore, it is ground into a powder and thoroughly mixed. This powder is then pressed into a pellet for analyzing purposes. The pellet is formed by first pressing the powder at 10 tons into a tablet about 26 mm in diameter. This is then placed in another mold 35 mm in diameter, surrounded and covered by boric acid, and pressed at 40 tons into a final double layer tablet. The boric acid backing adds strength to the pellet. For quantitative analysis of the sample, the counting rates of the single elements are compared with those of an appropriate standard whose composition is well known. NBS-standards are usually applied, especially granite G-1 and diabase W-1.

Typical experimental conditions for determining the counting rates of the potassium and calcium K-lines (K_α and K_β) are given in Fig. 6 and Fig. 7 using the standards G-1 and W-1 and a Philips X-ray fluorescence spectrometer (C. H. F. MÜLLER, Hamburg, Germany). G-1 and W-1 were prepared after the method of ROSE et al. (1963). The potassium K_α-intensity for W-1 amounts to about 500 cps corresponding to 0.54% K (Fig. 7). This illustrates the sensitivity of the X-ray fluorescence method which is suitable for low level potassium analysis.

Sources for IV.1.: BIRKS (1964), EWING (1960), SHALGOSKY (1960), HAHN-WEINHEIMER and ACKERMANN (1963).

2. Mass spectrometric isotope dilution analysis

The quantitative determination of potassium in materials with a potassium content lower than 0.2% is extremely difficult with gravimetric and other methods. Stable isotope dilution or neutron activation technique are more suitable for low level potassium analysis.

Isotope dilution analysis yields good results in potassium argon dating of stony meteorites in which the potassium content is almost always less than 0.1%. For the chondrites it is about 800 ppm, for the achondrites it ranges from 10 to 400 ppm. Mostly small quantities of meteorites are available, so that the absolute amount which has to be determined is only about 10^{-5} g potassium and lower. This method is used also in dating of terrestrial minerals with low potassium content (olivine, pyroxene, etc.).

For stable isotope dilution the quantity of an element is determined from the change produced in its isotopic composition by the addition of a known quantity of monoisotopic tracer of that element. The potassium tracer generally used is K^{41}.

The principles of this method are as follows: x grams potassium of normal isotopic composition with the three isotopes K^{39}, K^{40}, K^{41} and with the relative abundances N^{39}, N^{40}, N^{41} shall be determined. Then y grams of tracer potassium are added to x grams of normal potassium. The chemical form of normal potassium and tracer potassium is assumed to be different. The relative abundances of these three isotopes are assumed to be T^{39}, T^{40}, T^{41}. After bringing all of the potassium in a uniform chemical form, the isotopes in this compound have the relative abundances M^{39}, M^{40}, M^{41}. The atomic weights of normal and tracer potassium shall be m_N and m_T. The number of isotopes shall be $i = 39, 40, 41$. Then the following equation results:

$$\frac{x}{m_N} \cdot N^i + \frac{y}{m_T} \cdot T^i = \left(\frac{x}{m_N} + \frac{y}{m_T} \right) \cdot M^i .$$

We obtain for x:

$$x = y \cdot \frac{m_N}{m_T} \cdot \frac{M^i - T^i}{N^i - M^i}\ .$$

In the mass spectrometer no relative isotopic abundances but isotopic ratios are measured. Then it is more suitable to introduce isotopic ratios in the above equation.

With $K = 39, 40, 41$ one receives

$$x = y \cdot \frac{M^i/M^K - T^i/T^K}{N^i/N^K - M^i/M^K} \cdot \frac{m_N \cdot T^K}{m_T \cdot N^K}\ ,$$

$$x = y \cdot \frac{M^i/M^K - T^i/T^K}{N^i/N^K - M^i/M^K} \cdot \frac{\sum\limits_{i=1}^{i=n} N^i/N^K \cdot m_i}{\sum\limits_{i=1}^{i=n} T^i/T^K \cdot m_i}\ ,$$

m_i is the atomic weight of the isotope with the mass number i.

For the measurement the following conditions should be fulfilled:

(1) The relative abundances of the isotopes in the tracer should greatly differ from those in the element which is to be determined.

(2) For the isotopes in question no disturbing lines should exist.

(3) The relative abundance of the isotope in the normal element which is added as tracer, should not be small.

(4) The half-lives of radioactive isotopes must be long enough to prevent changes of isotopic ratios during the measurement.

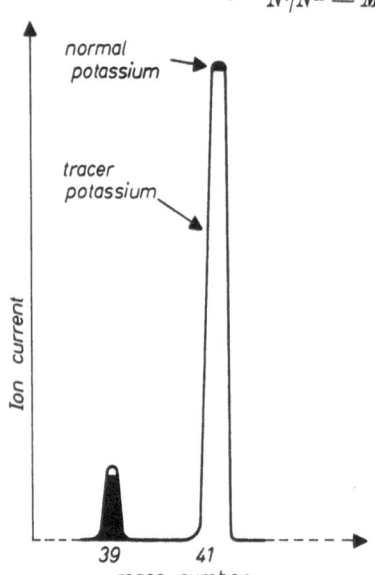

Fig. 8. Mass spectrum of potassium (from KRANKOWSKY, 1960), 2.7×10^{-8} g meteoritic (normal) and 2.0×10^{-5} g tracer potassium

These are fulfilled for K^{39} and K^{41}. Their abundances in the natural potassium are 93.08% and 6.91% respectively.

Fig. 8 shows the mass spectrum of meteoritic potassium and the contributions of meteoritic and tracer potassium to the spectrum.

The isotope dilution analysis is very sensitive for potassium: the detection limit is about 10^{-11} g potassium. The applicability of this method is limited by potassium contamination from chemicals used to treat the sample. Thus only highly pure chemicals may be used in isotope dilution.

Copious references for mass spectrometric isotope dilution analysis are given in "Methods in Geochemistry" by SMALES and WAGER (1960) and also in "Massenspektrometrische Kaliumanalysen an Steinmeteoriten nach dem Isotopenverdünnungsverfahren" by KRANKOWSKY (1960). (Partially published in KIRSTEN et al., 1963.)

3. Neutron activation analysis

Neutron activation: a radioactive isotope is produced by thermal neutrons from a stable isotope. The process for the activation of potassium is as follows:

$$K^{41} (n, \gamma) K^{42} \xrightarrow[12.4\,h]{\beta^-, \gamma} Ca^{42} .$$

This signifies that stable K^{41} forms by neutron capture the unstable K^{42} (half-life 12.4 h). K^{42} decays under β^- and γ emission to stable Ca^{42}.

Neutron activation is, for most elements, an ultra-sensitive method of trace analysis making it possible to measure concentrations of less than a few parts per billion. Furthermore this method is free from interference by other elements since the rate of decay and the radiation emitted are characteristic of the nuclide produced and will not be duplicated by any other radionuclide. Contrary to other analytical methods, once the irradiation has been finished, there is no danger of contamination.

The activity produced at a time t of the pile irradiation is

$$A = N \cdot \Phi \cdot \sigma (1 - e^{-\lambda t})$$

where A is the activity in disintegrations per second, N the number of atoms of the target nuclide, Φ the neutron flux per square centimeter per second, σ the activation cross section for the reaction in square centimeters per atom, and $(1 - e^{-\lambda t})$ is the saturation factor (the ratio of the quantity of activity produced in time t to that produced in infinite time). The decay constant λ and the half-life $T_{1/2}$ of the radionuclide are related by

$$\lambda = \frac{0.693}{T_{1/2}} .$$

The total number of atoms N of the target nuclide is related with the weight m of the target material as follows:

$$\frac{m}{M} = \frac{N}{N_L} ,$$

where M is the atomic weight and $N_L = 6.02 \times 10^{23}$ (LOSCHMIDT's number).

Then we obtain for the activity:

$$A = \frac{N_L \cdot \Phi \cdot \sigma \cdot m}{M} (1 - e^{-\lambda t}) .$$

If the element to be determined consists of different isotopes one must regard the natural abundance of the isotope activated, with the factor $f \leq 1$, in the equation above:

$$A = \frac{N_L \cdot \Phi \cdot \sigma \cdot m \cdot f}{M} \left(1 - e^{-\lambda t}\right).$$

Neutron activation method as an analytical tool: irradiating samples and a standard with thermal neutrons and comparing the activities produced. This comparative method eliminates the necessity of: (1) monitoring the neutron flux which is usually difficult to control, and (2) correction for absolute counting rates of the radioactive source. One has only to measure the activity of the element in the sample and the standard, and to calculate (with the known weight of the standard) the amount of the element in the sample:

$$\frac{\text{weight of element in sample}}{\text{activity of element in sample}} = \frac{\text{weight of standard}}{\text{activity of standard}}.$$

This calibration technique is, however, only applicable for small samples with neglectable neutron absorption. In practice, this condition is usually fulfilled.

After the irradiation, it is necessary to chemically separate the desired radionuclide from the other activated elements of the sample. This is done as follows: The specimen is decomposed. A known amount of the element investigated is added, as an inactive carrier, to the solutions of both the sample and the standard. The carrier element is then chemically separated from disturbing radionuclides. It is finally enriched (using a characteristic precipitation reaction), mounted on a planchet and counted. The activity of the radionuclide produced in the specimen is compared to that of the standard by suitable counting equipment. The purity of the activity is checked by measuring the half-life and the energy of radiation. Gamma-emitting nuclides are identified by using a scintillation detector with multi-channel spectrometer, β^- emitting radioisotopes are registered with Geiger or proportional counters. The energy of the radiation is controlled by using the range-energy relation (absorption technique).

During the decontamination procedure a quantitative recovery of the added tracer element cannot be achieved. Thus, the chemical yield both of specimen and standard must be determined, and activities corrected to 100%.

The method of neutron activation analysis is reviewed in many papers and bibliographies. A selection is cited here: BOYD (1949), GIBBONS et al. (1957), HERR (1960), MAPPER (1960), MEINKE (1958), MOORBATH (1960), SCHULZE (1962), SMALES (1949, 1953, 1956, 1957), TAYLOR and HAVENS (1956), WINCHESTER (1960).

V. Applications of potassium analysis to K-Ar dating
1. Terrestrial material

A preliminary mineralogical examination of the samples by thin sectioning and X-ray diffraction analysis is necessary to exclude doubtful material. One of the basic problems in K-Ar dating is considering the diffusion behaviour of both argon and potassium in the various potassium minerals (treated by FECHTIG and KALBITZER in this book).

Precision and accuracy are important factors in absolute dating of geological materials leading to ages which are stratigraphically useful.

The problem of potassium analysis in terrestrial materials shall be discussed in the following.

Several years ago a biotite sample, B-3203, was prepared as an inter-laboratory standard by FAIRBAIRN at M.I.T. Pinson (1961) compiled the data as shown in Table 5.

Table 5. *Analyses of biotite B-3203 for potassium*

Potassium (%)	Method	Analyst
7.80 ± 0.07	Neutron activation	J. WINCHESTER, M.I.T.*
7.40	Flame photometry, Li, internal standard	H. FAUL, United States Geological Survey, Washington, D.C.
7.42	Laurence-Smith: K_2PtCl_6	S. GOLDICH, University of Minnesota, Minneapolis, Minn.
7.79	Flame photometry, direct	L. KOVICH, University of California at Berkeley, Berkeley, Calif.
7.58 ± 0.03	Flame photometry, Li internal standard	W. PINSON, M.I.T.
7.40 ± 0.06	Gravimetric; $KClO_4$	W. PINSON, M.I.T.
7.71 ± 0.07	Flame photometry: Li internal standard	R. W. STOENNER, Brookhaven National Laboratory, Upton, N.Y.
7.68 ± 0.06	Laurence-Smith: K_2PtCl_6	R. W. STOENNER, Brookhaven National Laboratory, Upton, N.Y.
7.76	Flame photometry, Li internal standard	GEORGE EDWARDS, Shell Development Co.
7.59, 7.60	Isotope dilution	L. E. LONG, Lamont Geochemical Laboratory, Palisades, N.Y.
7.56	DU flame photometry	SYDNEY ABBEY, Canadian Geological Survey, University of Ontario, Ottawa, Canada

* Massachusetts Institute of Technology, Cambridge, Mass.

The potassium analyses of this standard range from 7.4 to 7.8%. There is a relative error of about 6%. These results were surprising and demonstrate the possibility of relatively large errors in potassium work.

Nevertheless, it is possible to obtain a precision better than 1% for potassium with flame photometry (COOPER, 1963), provided that certain disturbing influences are excluded. Therefore, this is the most usual method for potassium determination in geological materials used for potassium argon dating. Indeed, the potassium concentration of the sample should not be smaller than 0.2%.

The minerals and rocks almost always contain a large amount of silicates. These must first be decomposed. There are primarily two ways to make potassium water soluble: the J. Lawrence Smith and the Berzelius method.

J. Lawrence Smith method

This procedure consists of heating the finely ground sample together with ammonium chloride and calcium carbonate. Thereby, the alkalies are transferred into water soluble chlorides which can be leached off the sintered mass. A disadvantage of this method is the danger of alkali loss. For high quality analyses acid decomposition of the sample is preferred (see below). However, the Lawrence Smith method still finds application in the analysis of acid insoluble silicates as zircon, beryl, topaz.

The Berzelius method

The powdered sample (which contains some or all of the following elements as major constituents: Si, Al, Fe, Ca, Mg, Na, K) is decomposed by heating it with a mixture of sulphuric and hydrofluoric acid. Silicon is volatilized as silicon tetrafluoride and the remaining elements are left as sulphates. The excess sulphuric acid is fumed off and the residue is dissolved in water. If the decomposition is not complete, the acid treatment has to be repeated once more. Often perchloric acid is used instead of sulphuric acid because of higher volatility.

Potassium of the final salt solution can be measured

(1) directly in the flame photometer.

(2) Al, Fe, Ti, Ca are removed by ammonia and/or ammonium carbonate precipitation to exclude interferences. The filtrate, containing K, Na, partly Mg, and ammonium salts, is then used for the flame photometric determination of potassium.

COOPER (1963) published a comprehensive paper on flame photometry of potassium in geological materials obtaining good results (with a sodium buffer) which are in agreement with isotope dilution measurements (COMPSTON and VERNON, 1962).

See also the papers of DEAN (1960), HAVRE (1961), ABBEY and MAXWELL (1960), BAKER and GARTON (1961), POLUEKTOV (1961), EVERNDEN and RICHARDS (1962).

Several NBS and other interlaboratory standards of different chemical composition were analyzed for potassium by a flame photometric method recently developed in our laboratory (MÜLLER, 1966). A Zeiss PF 5 filter flame photometer and propane-air as fuel were used (Fa. Zeiss,

Oberkochen, Germany). The results are compared with those obtained from other laboratories and by other methods (Table 6).

Table 6. *Comparison of potassium analyses on terrestrial standards*

Standards		Recommended by other authors % K	Zeiss PF5 flame photo-meter. Chem. separation by $(NH_4)_2CO_3$ % K (f)
Sample No.	Sample		
NBS G-1 . . .	granite	4.52 D (a)	4.56
		4.61 FP (b)	
		4.65 ID (c)	
NBS W-1 . . .	diabase	0.53 D (a)	0.54_6
		0.54_4 FP (b)	
		0.54_1 ID (c)	
NBS No. 1 A. .	argillaceous limestone	0.59 D (d)	0.61
NBS No. 76 . .	burnt refractory	1.28 D (d)	1.25
NBS No. 77 . .	burnt refractory	1.75 D (d)	1.74
P-207	muscovite	8.58 D (e)	8.63
			8.59 ID
Bern 4 M . . .	muscovite	8.62 FP (e)	8.81
			8.84 ID

D: different methods (mean values), FP: flame photometry, ID: isotope dilution. (a): FLEISCHER and STEVENS (1962), (b): COOPER (1963), (c): COMPSTON and VERNON (1962), (d): mean values of the National Bureau of Standards, (e): LANPHERE and DALRYMPLE (1965), (f): MÜLLER (1966).

The silicate samples were decomposed with a mixture of hydrofluoric and perchloric acid. The dried residue was dissolved in distilled water and transmitted to a 100 ml volumetric flask. Then Fe, Al, etc. were precipitated out of the hot solution with 5 ml of a 10% ammonium carbonate solution. The filtrate containing mainly K, Na, and excess ammonium carbonate was measured in the flame photometer. As a control sample, we used a solution: the concentration of which being similar to that of the sample relative to potassium and ammonium carbonate.

The excess of ammonium carbonate eliminates differences of physical properties of both sample and control solution (viscosity etc.).

The influence of cations remaining after the precipitation with $(NH_4)_2CO_3$ in solution was observed as follows:

Sodium only affects an enhancement of potassium if the Na/K ratio is greater than about seven.

Magnesium influences potassium only to a small extent, if aluminum and/or iron are present in the sample. It seems that the hydroxides of Al and Fe hold back magnesium.

Calcium does not interfere with the potassium emission even if the calcium concentration is high, because Ca is precipitated as carbonate.

Summarizing, we can say that flame photometry furnishes good results for potassium determinations in geological materials. Indeed, the potassium content of the samples should not be lower than 0.2%. The precision and accuracy of flame photometry for potassium is about ±1.5%.

2. Tektites

Tektites are natural glasses the origin of which is still enigmatic and under intensive discussion. Potassium argon dating of tektites is of great interest because in recent years age relations between them and impact glasses were found favoring a cogenetic terrestrial origin (GENTNER et al., 1961, 1963; GENTNER and ZÄHRINGER, 1960; ZÄHRINGER, 1963a; GENTNER et al., 1964).

The potassium concentration of tektites for various regions range between 0.8 and 3.6% K_2O. Flame photometry is thus recommended as a reliable and rapid method for potassium determination.

The crushed tektite sample containing 70—80% SiO_2 is decomposed with a mixture of hydrofluoric and perchloric acid. The dried residue is dissolved in distilled water and transferred to a volumetric flask. Al, Fe, Ca, Ti are precipitated as described in section V.1). Potassium is measured with a filter flame photometer. The procedure (MÜLLER, 1966) was checked with the philippinite standard, Po-300 (received by the courtesy of Dr. F. CUTTITTA) prepared at the Geological Survey, Washington D.C., and with the opal glass, No. 91, of the NBS. Table 7 shows that the values are in good agreement to those recommended:

Table 7. *Comparison of flame photometric results for potassium in two glass standards*

Standards	% K	
Philippinite tektite Po-300 .	2.00	GS
	1.95	MÜLLER
Opal Glass No. 91, NBS . .	2.70	NBS
	2.77	COOPER
	2.74	MÜLLER

Analysts: GS = Geological Survey, Washington D.C. NBS = National Bureau of Standards, Washington D.C., COOPER (1963), MÜLLER (1966).

Detailed information on tektites and their K-Ar ages is given in this book by SCHAEFFER.

3. Stony meteorites

Stony meteorites are classified in two main groups: chondrites and achondrites. This grouping is also expressed in the potassium content: chondrites have an average potassium content of about 900 ppm K, while the content of achondrites is lower and varies considerably between 10 and 400 ppm K.

It is obvious that classical chemical methods and flame photometry fail in the potassium determination. Good results are achieved with the isotope dilution method and the neutron activation technique. In the following isotope dilution procedure is discussed.

The selection and the treatment of the specimen requires great care since terrestrial potassium can easily contaminate the meteoritic sample. In addition, only ultra pure chemicals should be used for the chemical treatment. A further difficulty lies in taking a representative portion of the meteorite since meteorites are heterogenious bodies having besides silicates, a metallic and a troilite phase. It is advisable to start the analysis with at least 100 to 200 mg of the specimen.

The isotope dilution technique for potassium was applied to a large number of stony meteorites yielding reliable results (KIRSTEN et al., 1963).

The sample is decomposed with a mixture of hydrofluoric and hydrochloric or perchloric acid after adding tracer potassium in an amount which corresponds to the meteoritic potassium. The residue is dissolved in water and Al, Ca, Fe, etc. are precipitated with ammonium carbonate. The filtrate is evaporated and the dry salts are heated up to 600° C. The potassium is leached out of the sintered mass with water. The potassium solution is separated by filtration from the residue, and transferred to the filament, which is mounted in the mass spectrometer.

As mentioned above the purity of the water and the chemicals used is an important precondition.

In the potassium measurements reported by KIRSTEN et al. (1963) the blanks of the chemicals were as follows (measured by isotope dilution technique):

$$H_2O \quad : \quad 3 \cdot 10^{-8} \text{ g K/g}$$
$$HF \quad : (1.4 \pm 1) \cdot 10^{-7} \text{ g K/g}$$
$$HCl \quad : (1.2 \pm 0.2) \cdot 10^{-7} \text{ g K/g}$$
$$(NH_4)_2CO_3 : (1.1 \pm 0.2) \cdot 10^{-7} \text{ g K/g.}$$

HF, HCl, $(NH_4)_2CO_3$ were available in this purity (Fa. Merck, Darmstadt). Tap water was first distilled in a metal apparatus and then deionized in a mixed bed ion exchange column.

Recently these blanks were improved (MÜLLER, 1965) by redistilling the deionized water in a quartz column. We got a potassium concentration of $2 \cdot 10^{-9}$ g K/g. This water served as basis for the production of hydrochloric acid, ammonia and ammonium carbonate. These were produced by bubbling the corresponding bomb gases into the high pure water.

By this method we got the following potassium concentrations:

HCl (13 molar): $6 \cdot 10^{-9}$ g K/g

NH$_3$ (12 molar): $8 \cdot 10^{-9}$ g K/g

$(NH_4)_2CO_3$ (saturated solution): $1.2 \cdot 10^{-9}$ g K/g.

The low value of $(NH_4)_2CO_3$ is surprising: it seems that potassium is enriched in the $(NH_4)_2CO_3$-crystals. The isotope dilution measurements were made by KAISER (1965).

4. Iron meteorites

Potassium argon dating of iron meteorites is difficult since the potassium concentrations of these objects range between 10^{-7} and 10^{-9} g K/g. K-Ar dating attempts were undertaken by STOENNER and ZÄHRINGER (1958). Their experiments were continued by MÜLLER and ZÄHRINGER (1966). Independently, FISHER (1965) tried to date some iron meteorites.

The highly sensitive neutron activation technique was applied for both the potassium and the argon determination by the authors cited above, using the processes K^{41} (n, γ) K^{42} and Ar^{40} $(n, (\gamma)$ Ar^{41}.

The procedure used by MÜLLER and ZÄHRINGER 1966) is described briefly:

A compact 10 g piece of iron meteorite was sealed off in a quartz tube under a vacuum. It was then irradiated for one hour in the Karlsruher reactor FR 2 at a neutron flux of about 3×10^{13} n/cm² sec. After the irradiation, the surface of the sample was cleaned mechanically by removing about 10 % of the surface material and washed with aceton and water to exclude surface contamination by terrestrial potassium. For the determination of the argon the specimen was transferred to a quartz oven with molybdenum crucible. The gas was extracted in several steps by heating and melting. The gas was purified together with an inactive carrier of argon in the usual manner and transferred with a Toepler pump to a Geiger counter. The radiochemical purity of the Ar^{41} was confirmed by measuring the half-life to be 1.8 hours.

For calibration the counting rate of Ar^{41} was related to the total amount of Ar^{40} by irradiating in the same run a measured quantity of Ar^{40} and a known amount of potassium (as potassium acid phthalate) as flux monitor (see below). The specific activity of Ar^{41} resulting from a standard irradiation is analogous to the specific activity of K^{42} resulting from the same irradiation. Under our conditions 1 dpm was equivalent to about $2 \cdot 10^{-10}$ cm³ of Ar^{40}.

Regarding side reactions: a further source of the Ar^{41} can be K^{41} produced by the reaction K^{41} (n, p) Ar^{41} with fast neutrons. Thus, KCl was irradiated under the standard conditions with the result that 10^{-7} g of potassium gave about 1 dpm of Ar^{41}. In addition, Ca^{44} produces Ar^{41}

by a (n, α) reaction. However, irradiation of $CaCO_3$ showed only an Ar^{37} activity by a (n, α) reaction on Ca^{40}.

Air contamination of the meteorite sample could be seen by degassing the specimen in several steps at different temperatures (400° and 800° C) and counting the Ar^{41} and Ar^{37} of the single fractions.

For the potassium determination the reaction K^{41} (n, γ) K^{42} was used. Experiments showed that the potassium in the iron meteorite was volatile under the conditions of the argon extraction. This means that potassium evaporates onto the furnace walls. The potassium content of the residue remaining after the melting of the sample in the crucible, was always below 1 % of the total potassium.

The potassium was washed from the furnace walls with a mixture of concentrated HCl, HNO_3, and HF. Inactive potassium carrier (30 mg K) was added, and the acid solution well mixed. After the extraction of iron with di-isopropyl ether, the potassium was precipitated with sodium tetraphenylboron. Decontamination steps of hydroxide and basic carbonate precipitations and an anion exchange step for iron (from 8n HCl on Dowex 1) followed using different inactive carriers. Potassium was precipitated again with sodium tetraphenylboron, then transferred in the perchlorate, recycled and finally mounted on an aluminum planchet and counted. The samples were counted with a two inch end window beta proportional counter, and the counting rates were compared directly to the counting rate of the potassium acid phthalate standard.

Calculated for the end of the irradiation 1 dpm corresponds to about $5 \cdot 10^{-11}$ g of potassium.

The radiochemical purity of the K^{42} was checked by measuring the half-life to be 12.4 hours, and determining the range-energy relation of the β^- radiation by absorber technique.

Finally, for sample and standard the chemical yields were determined by flame photometry. Counting rates were corrected for 100%.

The present method for potassium gives an accuracy of $\pm 5\%$, and confirms the error given by STOENNER and ZÄHRINGER (1958) for the potassium activation analysis of the stony meteorite, Forest City (test sample).

The activation analysis of potassium was also applied with success to stony meteorites: it is useful in analyzing separated mineral fractions as was shown for the Indarch meteorite by SCHAEFFER et al. (1965).

A further possibility to determine potassium by activation analysis is given by the K^{39} (n, p) Ar^{39} reaction using a reactor with fast neutrons. This method is applicable on samples with low Ca/K ratios only, because of (n, α) reactions on Ca. The sample has to be wrapped in cadmium foils to avoid thermal neutrons. It has first been tried by ZÄHRINGER and FIREMAN (1956) on iron meteorites. Later WÄNKE (1961) determined the potassium content of some stony meteorites with this technique.

The Diffusion
of Argon in Potassium-Bearing Solids

H. Fechtig and S. Kalbitzer

1. Introduction

The diffusion of rare gases in solids was first investigated by Ruther-ford (1901) and Strutt (1909). More intensive work was performed in the 1930's by Hahn and his coworkers — Cook (1939), Flügge and Ziemens (1939).

A specific case, the diffusion of argon in solids, became important during the development of the K-Ar dating method. This occured because the K-Ar dating of certain minerals gave lower values than was obtained by geological methods. The K-Ar procedure, however, did agree with geological methods for other minerals, mica being one.

The diffusion effect was noted in 1950 when Smits and Gentner tried to date the salt mine at Buggingen, Baden, Germany. They found that large salt grains contained more argon per unit weight than small grains — a typical diffusion effect (Gentner, Präg and Smits, 1953 a, b).

About the same time, many groups were at work classifying the most important minerals as to their ability to retain radiogenic argon. Gerling (1939) pointed out that the activation energy of the diffusion of argon measured at high temperatures should be a criterion for the classification. However, since transport phenomena in solids are usually given by more than one activation energy, it cannot be concluded that the same mechanism operating at high temperatures will also operate at room temperature. In the last few years, our group has carried out diffusion measurements at low temperatures in order to decide whether diffusion losses must be taken into account.

2. Principles of diffusion

2.1. Fundamental laws

Diffusion is an irreversible process smoothing out concentration differences in a system by a corresponding flow of particles.

The first and second Fick's Law are given by:

$$J = -D\nabla c, \tag{1}$$

$$dc/dt = \nabla(D\nabla c). \tag{2}$$

J denotes the diffusion current, which is proportional to the concentration gradient ∇c. The diffusion constant D is a measure of the ease of the equalization. It is temperature dependent and reflects the structural properties of the substance under consideration. The second equation describes the change of concentration with time. Usually the dependence of D on c can be neglected to give:

$$dc/dt = D\Delta c . \tag{3}$$

Eq. 1 is used for steady state experiments ($\nabla c = \text{const.}$). In our case the concentration of argon is time dependent. Therefore we will exclusively deal with the solution of equation (3).

2.2. Boundary conditions and initial values

In the case of argon diffusing out of a mineral we are dealing with a finite system characterized by certain restrictions. These restrictions are concerned with the concentration of argon at the boundaries of the grains and with the initial conditions.

Starting with a simple model, we neglect the actual increase in time of argon concentration by radioactive decay. This will be taken into consideration in chapter 3. Consider a sphere with a radius R and a homogenous distribution of argon at time zero. Furthermore, it shall be assumed that an argon concentration in the space outside the sphere,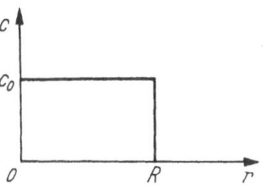

Fig. 1. The boundary conditions at time zero

$r > R$, gives no rise to a backdiffusion into the sphere. This assumption is valid even for relatively high outer concentrations, if the potential barrier for particles to enter the crystal is very high. The space outside the sphere can therefore be treated as having a concentration zero at all times. In Fig. 1 this situation is summarized.

For $t = 0$: $c = c_0,$ $0 \leqq r < R$

$c = 0,$ $R < r .$

For $t > 0$: $c = c(r, t),$ $0 \leqq r < R$

$c = 0,$ $R \leqq r .$

This is a complete description of our system; the exact solution of which is derived by CARSLAW and JAEGER (1959).

$$F = 1 - \frac{6}{\pi^2} \sum_{n=1}^{\infty} \frac{1}{n^2} e^{-n^2 Bt}; \quad B = \frac{\pi^2}{R^2} D . \tag{4}$$

Here F, the fractional loss of argon, is given as a function of time.

2.3. Practical mathematical forms

Equation (4) is only of limited use since the convergence of the infinite series is only good in case of high losses; for low and moderate losses other forms were derived by Carslaw and Jaeger (1959) and tabulated by Reichenberg (1953). For small losses, $F \leq 10\%$:

$$F = \frac{6}{\pi^{3/2}} \sqrt{Bt}; \quad D = \frac{F^2 \pi R^2}{36t} . \tag{4a}$$

For losses $F < 90\%$:

$$F = \frac{6}{\pi^{3/2}} \sqrt{Bt} - \frac{3}{\pi^2} Bt;$$

$$D = \frac{R^2}{\pi^2 t} \left(2\pi - \frac{\pi^2}{3} F - 2\pi \sqrt{1 - \frac{\pi}{3} F} \right). \tag{4b}$$

For high losses, $F > 90\%$:

$$F = 1 - \frac{6}{\pi^2} e^{-Bt} ;$$

$$D = - \frac{R^2}{\pi^2 t} \ln \left\{ \frac{\pi^2}{6} (1 - F) \right\}. \tag{4c}$$

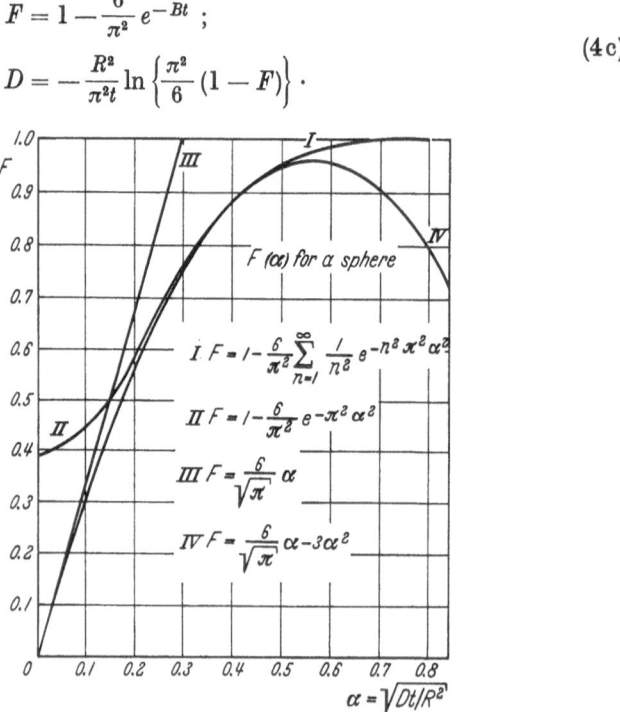

Fig. 2. Fractional release for spherical crystals according to the exact (I) and some approximate (II—IV) solutions of the diffusion equation

In Fig. 2 the exact solution (4) and the approximations are shown (after Lagerwall and Zimen, 1964).

Therefore, if the total concentration c_0 of argon has been determined — e.g. by melting a second sample or by summing up all losses during the heating and adding the remainder — the diffusion constant is readily calculated from the heating time t and the corresponding fractional loss F.

It is noteworthy, that the calculation is exactly correct only when the concentration at the beginning of the run was ideally square. This is of course never the case.

The disturbance of the square profile can only be neglected when the fraction released by the experiment is several times higher than the original loss; in this case the result-ing diffusion coefficient will be right.

In particular, if a sample with argon losses (e.g. a mineral to be dated) is taken for diffusion experi-ments, one should be aware of the violation of the boundary conditions. The apparent diffusion constants will come out too low, and the activation energies too high. It is, however, possible to correct for this, e.g. by a determination of the true concen-tration profile (sectioning methods) or by a careful investigation of the kinetics. A similar problem arises, if

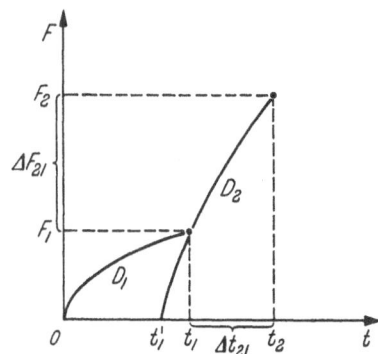

Fig. 3. Schematic illustration of mathematical treatment correcting for the disturbance of the profile after the first run. The second run is performed at a higher temperature

a sample is to be used at different temperatures. After the first run the profile is similarly disturbed; but from the knowledge of all losses the following forms can be used, (see Fig. 3) analogous to eq. 4a, 4b, 4c, permitting a correct evaluation:

$$D_{i+1} = \frac{(F_{i+1}^2 - F_i^2)\,\pi R^2}{36 \cdot (t_{i+1} - t_i)}\; ; \tag{5a}$$

$$D_{i+1} = \frac{R^2}{\pi^2(t_{i+1} - t_i)} \left\{ -\frac{\pi^2}{3}\,(F_{i+1} - F_i) \right.$$
$$\left. - 2\pi \left(\sqrt{1 - \frac{\pi}{3}\,F_{i+1}} - \sqrt{1 - \frac{\pi}{3}\,F_i} \right) \right\}\; ; \tag{5b}$$

$$D_{i+1} = \frac{R^2}{\pi^2(t_{i+1} - t_i)} \ln\left(\frac{1 - F_i}{1 - F_{i+1}} \right). \tag{5c}$$

The index i gives the number of the run of the sample, which is charac-terized by a diffusion time t_i, by a loss F_i and a diffusion constant D_i. If the next run is performed at a different temperature, the diffusion coefficient D_{i+1} will be found by inserting the corresponding values for t and F. F always means the sum of all previous losses.

In the low temperature range the released fraction F_{i+1} usually exceeds F_i by only a small amount; this gives

$$D_{i+1} = \frac{R^2 \pi F_i \Delta F_{i+1,i}}{18 \Delta t_{i+1,i}}. \tag{6a}$$

Although the one dimensional problem of a sphere can be treated most easily, the actual shapes of naturally occurring crystals are better described by other geometric shapes like cubes or parallelepipedes.

Lagerwall and Zimen (1964) published tables to be used for these more specific models.

2.4. Irregularly shaped crystals

Usually, age determinations are carried through with natural crystals having irregular structures for which spherical approximations may cause significant errors. In treating this case Wrage (1962) used a method similar to that of Pohlhausen (1921), where only the surface region of the crystal is taken into account. This is a good approximation, as long as the depletion layer stays within a critical distance from the surface, this distance being small in comparison with the smallest dimension of the crystal. Generally, the approximation is good to losses up to 30%.

Wrage's equation reads:

$$D = \frac{L^2 F^2}{12t},$$ (7)

where $L = 3 \times \frac{\text{volume}}{\text{surface area}}$, $t = $ time, $F = $ fractional loss.

In this range of loss the surface to volume ratio rather than the particular shape of the sample is important. Keeping this in mind, the radius of the sphere is replaced by a generalized length, L, in eq. (4a).

2.5. Samples containing a variety of grain sizes

A further complication of the conditions is brought about by the fact that natural samples usually contain different grain sizes. Sometimes the interval of sizes can be limited by sieving. Frequently, however, this is not possible, when the grains are of the dimensions of microns or when the sieved fractions have a different true grain size. It was shown by Fechtig, Gentner and Lämmerzahl (1963), that the application of the simple formulas on this latter case yields erroneous diffusion constants, as soon as the smallest grains have lost appreciable amounts of argon. If these smaller grains comprise a considerable fraction of the total weight, the calculated diffusion constants and the activation energies will be too low.

It is possible to correct for this effect experimentally, since calculations are no longer indicative. Several samples have to be degassed at different temperatures. The true diffusion coefficients are then determined by extrapolation to low losses. The small activation energies obtained for the argon diffusion in stony meteorites by Goles, Fish and Anders

(1960) are probably due to this effect. FECHTIG et al. (1963) found activation energies of 40—50 kcal/mol, which are in the range of the usually observed energies for the diffusion of argon.

Similarly complicated conditions are met in the diffusion of radioactive fission products out of powdered or sintered nuclear fuels.

MILLER (1960) treated this case under the assumption, that losses could be kept so small that the depletion layer developing from the surface was still small to smallest crystal dimension. Then in eq. (4a) the ratio of the total surface area, A, to the total volume, V, of the sample replaces the radius of the sphere:

$$F = \frac{2}{\sqrt{\pi}} \frac{A}{V} \sqrt{Dt} \ . \tag{8}$$

If the range of grain sizes cannot be limited, one has to resort to the experimental procedure described above.

2.6. Samples of complex composition

Often the samples are composites of a number of different minerals. The separation of the different components may be impeded by the fact that only small amounts of the material are available, or that the figure of separation is low. As long as there is not a large variation in grain size, it should be possible to distinguish the different activation energies of each component.

This situation is somewhat analogous to the problem, that inside one mineral the argon can assume different lattice positions, which are defined at respective critical temperatures. For these complex systems the linear heating technique may prove to be a powerful tool of investigation.

2.7. Activation energies

Let us consider the one-dimensional scheme of a lattice, Fig. 4. To reach the next stable position the argon atom has to overcome the potential barrier of height E[1]. Vibrating in the potential well it statistically gains enough energy

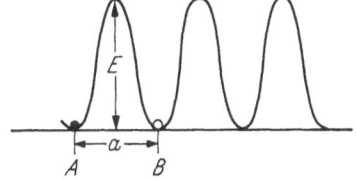

Fig. 4. Schematic course of lattice potential in a crystal showing the barrier between stable positions at A and B

to surmount the barrier. The experimentally observed temperature dependence of the diffusion constant is given by the Arrhenius equation:

$$D = D_0 \, e^{-Q/RT};$$

Q is the activation energy; R, the gas constant; T, the absolute temperature; D_0, a characteristic constant.

---

[1] See e.g. SEEGER, A. (1955).

In general Q may consist of different components; D_0 may contain contributions from thermodynamic potentials besides a frequency term, the lattice parameter and a geometrical factor[1]. Usually natural systems are characterized by several activation energies for argon diffusion. As mentioned above this can be caused by the argon atoms assuming different lattice positions. Competing transport mechanisms could also account for different activation energies. When structural defects (vacancies etc.) are involved in the process, the temperature pattern is usually not simple.

2.8. Non-volumic diffusion

Sometimes fractions of weakly bound argon are found in diffusion experiments. At lower temperatures a flat curve in the Arrhenius plot is observed, characterized by rather low activation energies. In substances where the potassium is not a regular lattice ion one might suspect that radiogenic argon will be located at grain boundaries and consequently be bound only loosely. In crystals, where the potassium is a regular constituent, the existence of loosely bound argon may be explained by higher structural defects (such as dislocations, grain boundaries, etc.), in which case the argon will diffuse out easily. Even for synthetic single crystals of KCl such fractions, though very small, have been observed (see chapter 5).

3. Correction of ages

The differential equation describing the process of generation and diffusion of argon is:

$$\frac{dc}{dt} = D\Delta c + \lambda \frac{R}{1+R} K_0 e^{-\lambda t}, \qquad (10)$$

where R is the branching ratio of the decay for K^{40}, λ the decay constant and K_0 the initial number of K^{40} per cc. The other symbols have the usual meaning. The diffusion problems we dealt with before, were characterized by a maximum concentration $c = c_0$ at time $t = 0$. The minerals for K-Ar dating contain initially no argon and for this case Wasserburg (1954) and Nicolaysen (1957) obtained the following solution for spherical geometry:

$$\frac{Ar^{40}}{K^{40}} = \frac{R}{1+R} \left\{ -1 + \frac{3}{d^2} - \frac{3\cot d}{d} + \frac{6d^2}{\pi^2} \sum_{n=1}^{\infty} \frac{e^{\lambda t(1-n^2\pi^2/d^2)}}{n^2(d^2 - n^2\pi^2)} \right\} \qquad (11)$$

where $d^2 = \lambda a^2/D$; a is the radius of the sphere.

Tables relating the measured age and the true age were calculated from this relation. The parameter is the dimensionless quantity d^2 resp. d. The younger the material, the more negligible are the losses for

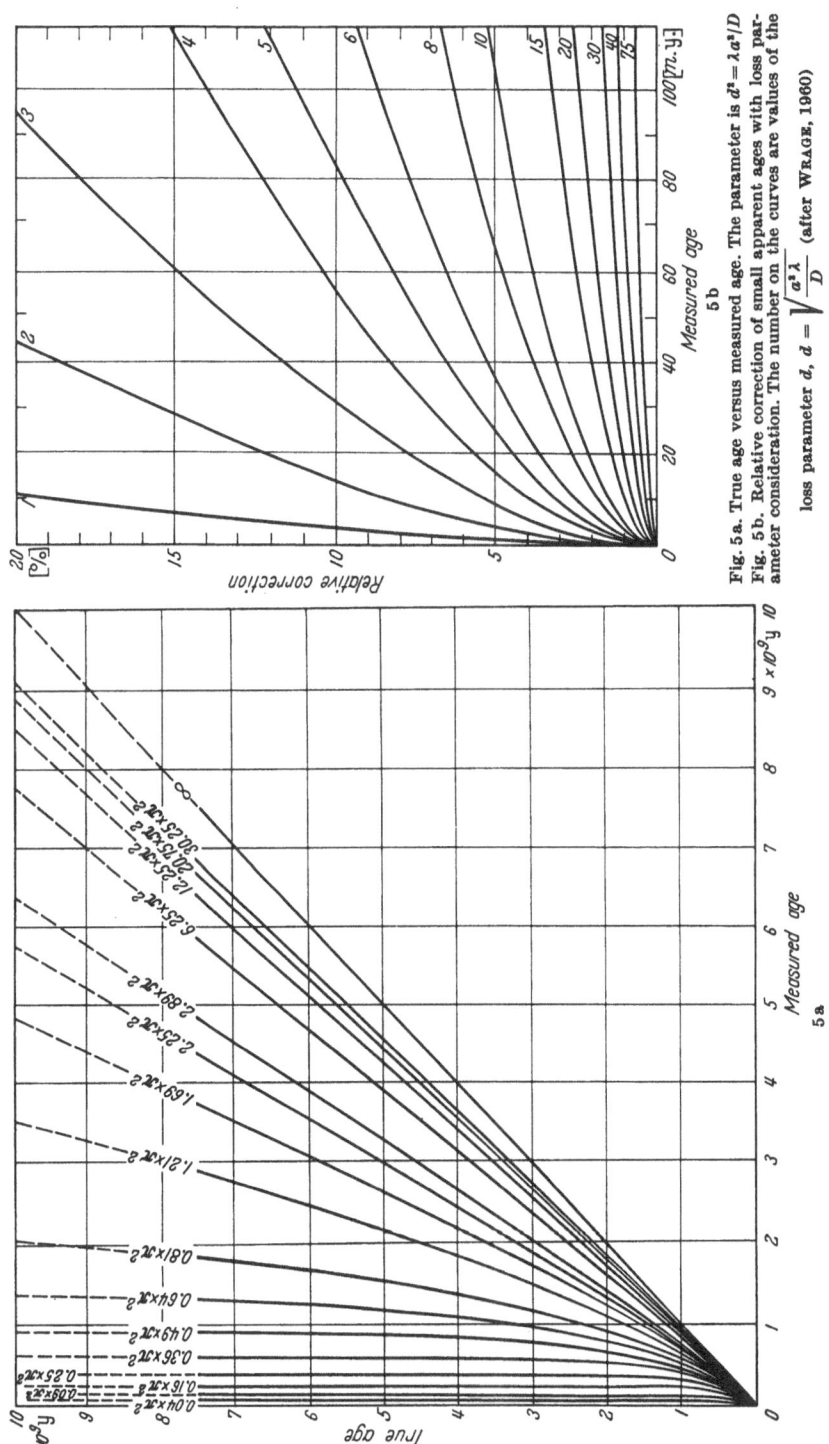

Fig. 5a. True age versus measured age. The parameter is $d^2 = \lambda a^2/D$
Fig. 5b. Relative correction of small apparent ages with loss parameter consideration. The number on the curves are values of the loss parameter d, $d = \sqrt{\dfrac{a^2\lambda}{D}}$ (after WRAGE, 1960)

high values of d. This can be seen in Fig. 5a and 5b. It is important that only the ratio of D to a^2 enters into eq. (11). As mentioned above, the true grain size of a natural sample is frequently an ill defined quantity. However, since the expression D/a^2 can be derived from diffusion measurements, one does not need to know D or a^2. Although the idea of correcting apparent ages in order to obtain true ages seems to be promising, it has only a limited practical value.

First, the diffusion constants are known no better than by a power of ten. In addition, a temperature history has to be assumed. When the activation energy is 50 kcal/mole, an error of 20 degrees in the lower temperature range means an error of more than two powers of ten in D.

Essentially, with regard to the figures 5a, 5b, diffusion measurements indicate whether diffusion took place or not. Correction, however, seems to be impossible under these conditions. Nevertheless, one can divide substances into two categories: (1) those being reliable and (2) those not appropriate for the K-Ar method.

If — as in the case of the salt mine of Buggingen — measurements of the content of argon as a function of grain size have been made, a better possibility of corrections for low ages and losses less than 30% is given. One finds after Wrage (1962) and Lippolt et al. (1963):

$$t = \frac{L_1 t_1 - L_2 t_2}{L_1 - L_2} \tag{12}$$

where t is the true age, t_i the apparent ages of different grain sizes and L_i the corresponding volume to surface ratios: $L_i = 3 \, V/A$.

Thus, from the results of Gentner, Präg, Smits (1953a, b), Lippolt et al. (1963) found the age to be approximately $19 \times 10^6 \, a$.

Thus no diffusion measurements are needed. The determination of the L_i may be quite elaborate, however.

4. Experimental methods

4.1. Mass spectrometry

When the diffusion measurements are done with radiogenic Ar^{40}, one uses high sensitive mass spectrometers. (As to the operation and construction, the reader is referred to the preceding article.) For the sake of time, several diffusion experiments are usually simultaneously carried out at different temperatures. This is especially true at low temperatures where measurable quantities do not diffuse out for several days.

Due to the high sensitivity of the spectrometers it is possible to measure diffusion constants at temperatures as low as about 200° C. Let us assume a background corresponding to 10^{-9} cc/STP Ar^{40}. Then from a sample with 10^{-5} cc the 10^{-3} part can be measured. This corresponds to a product of 10^{-11} cm^2 in Dt for a grain size of 10^{-2} cm. Heating for two days makes it possible to measure a diffusion constant of 10^{-16} cm^{2}sec^{-1}.

4.2. Neutron activation

Irradiation with neutrons is a convenient way to introduce radioactive isotopes of rare gases into a sample. In this case the sensitivity is greatly enhanced, since a few dpm can be detected by standard counting techniques. If the cross section of the reaction is not too low and the

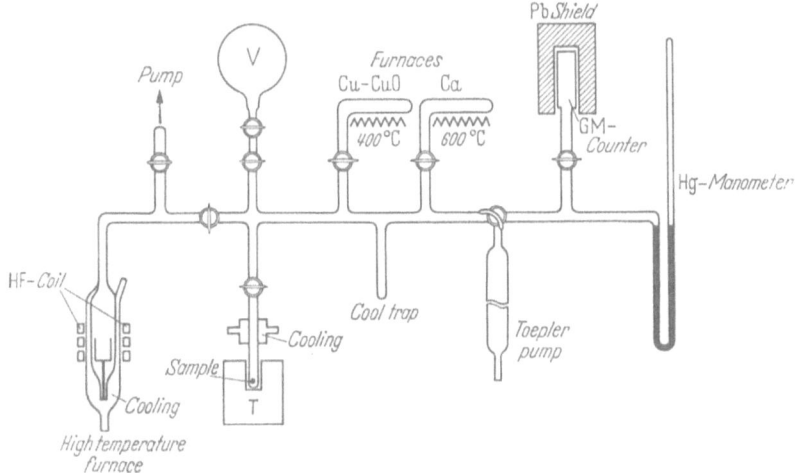

Fig. 6. Rare gas extraction system

half life of the isotope is not too long, then much lower diffusion constants can be measured than with a mass spectrometer. Fig. 6 shows a schematic representation of an apparatus used for diffusion experiments with rare gas isotopes. A small separate vessel containing the sample and some carrier (e.g. 1 Torr argon) is connected to the system. After cleaning for 10 to 30 min. over hot Ca — or Ti — ovens the remaining gas is put into the 4π-GM counter by means of a Toepler pump. For total extraction the sample is melted in a HF-heated furnace.

In the case of diffusion experiments on K-minerals argon is usually introduced in the neighborhood of potassium. This means that chiefly the reactions

$$K^{41} \quad (n, p) \quad Ar^{41} ,$$

$$K^{39} \quad (n, p) \quad Ar^{39} ,$$

have to be used.

The first reaction introduces a short lived isotope, $(T^{1/2} = 1.8 \text{ hours})$. This means that one is restricted to work close to the reactor and that only relative short diffusion times can be employed. A further inconvenience is the exact record of time, which is needed for the correction due to the decay. The Ar^{39} method is less sensitive because the half life of approximately $300a$ is so long.

Here high n-doses of the order of 10^{18} are necessary to introduce a sufficient number of atoms into the sample. This technique is used to work at lower temperatures. Since these high doses will strongly damage the lattice, one has to check for possible influences on the diffusion. On the other hand, the advantage is that the sample can cool down and immediate measurements are not necessary.

In minerals, where both calcium and potassium belong to the regular lattice one can take advantage of using calcium reactions. The process:

$$Ca^{40} (n, \alpha) Ar^{37}$$

introduces Ar^{37} which has a half life of 34 days. This method combines the advantages of the other methods; irradiation doses need not be very high, the samples can cool down and one is not restricted in time. Here even lower diffusion constants can be measured. A short estimate of the range of the methods is in order. The following table gives some data on the production of the isotopes by neutron irradiation[1]:

Isotope	Dose	Activity
Ar^{39}	$10^{19}n$	0.2 mC/gK
Ar^{37}	$10^{19}n$	25 mC/gCa

From eq. (4a) it follows with a grain radius of 10^{-2} cm, two days diffusion time, and a detection limit of 10 dpm, that, per-gram-potassium, a diffusion constant of 10^{-22} cm^{2}sec^{-1} is the limit. For per-gram-calcium the value is 10^{-26} cm^{2}sec^{-1}. With the latter method (Fechtig, Gentner. and Zähringer, 1960; Fechtig, Gentner, and Kalbitzer, 1961) it was actually possible to measure gas losses down to room temperature.

4.3. Isothermal heating

The isothermal heating technique allows the experimenter to control the kinetics of the diffusion. The released gas is measured continually or in appropriate intervals of time. To cover the entire temperature range of interest a number of samples are heated at fixed temperatures. If the kinetics given by eq. (4a, b, c), are obeyed, volume diffusion takes place. If there is more than one component of argon present, one should observe deviations from the ideal case. At certain temperatures it will be possible to determine the amounts of the different fractions.

---

[1] The irradiations were performed in the reactor of the TH München at Garching.

It should be noted, if the components diffuse independently, i.e. if one process does not feed another one, then the corresponding fractions have to be inserted for calculation of the diffusion constant.

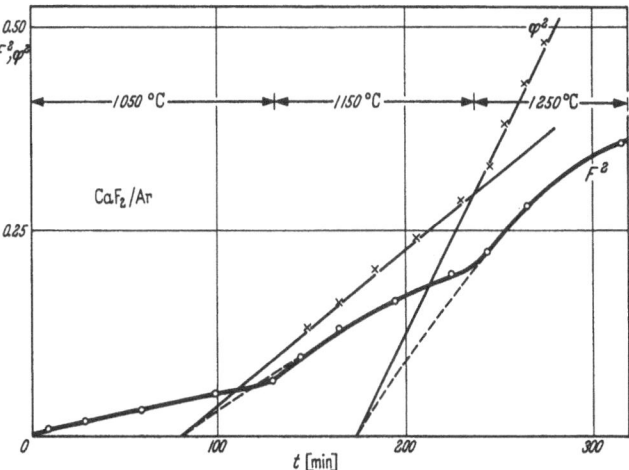

Fig. 7a. Diffusion of Ar³⁷ in a CaF₂ single-crystal (10 × 10 × 5 mm)

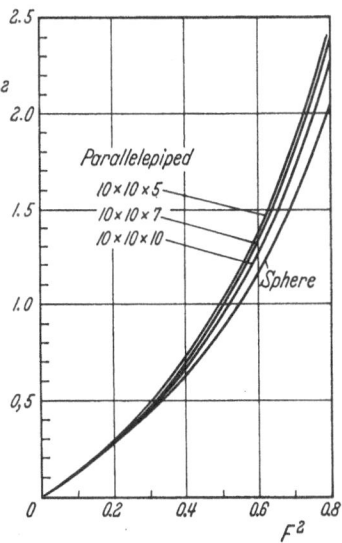

Fig. 7b. Relation between φ^2 and F^2

In case of small fractional release, F, it is convenient to plot F^2 versus time. In case of ideal diffusion a straight line will result. If losses have occurred, comparable to the amount released at that particular temperature, then the straight line should not go through the origin. The previously lost amount will be given by the ordinate at time zero. As MILLER (1960) points out, short preheating does not change the evaluation, since the diffusion constant is simply given by the slope of the straight line. For higher losses it is convenient to plot a quantity φ^2 instead of F^2 correcting for the second term in eq. (4b). FELIX et al. (1964) used this variable to investigate the diffusion of Ar³⁷ in CaF₂ (Fig. 7a); they give the relation

$$\varphi^2 = \frac{4}{\pi}\left(\frac{S}{V}\right)^2 Dt, \qquad (13)$$

where S/V is the surface to volume ratio of the specimen. Correction factors for the transformation of F^2 into φ^2 are shown in Fig. 7b.

4.4. Linear heating

Activation energies and different components of the argon diffusion can be determined by a method, which has been used in solid state physics to study the annealing of radiation induced defects (Vand, 1943; Parkins, Dienes, and Brown, 1951; Dienes and Vineyard, 1957). The sample is heated linearly with time. The resulting maxima of the rate of annealing and the heating rate are related to the activation energies.

Gerling et al. (1963) and Levskii (1963) applied this method to study the diffusion of rare gases in minerals and meteorites.

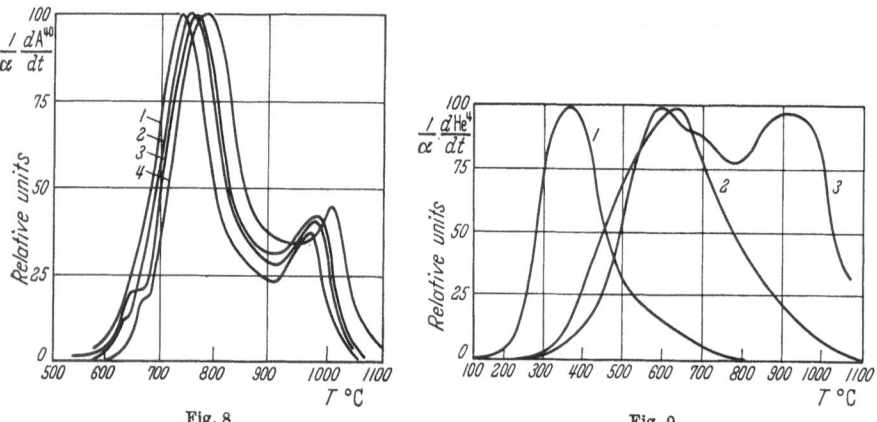

Fig. 8   Fig. 9

Fig. 8. Liberation of radiogenic argon from biotite (N. Karelia). Curve 1, $\alpha = 4.33$ deg/min, Curve 2, $\alpha = 5.58$ deg/min, Curve 3, $\alpha = 7.43$ deg/min, Curve 4, $\alpha = 11.0$ deg/min

Fig. 9. Staroye Pes'yanoye meteorite—curve 1, Zhovtnevyi Khutor meteorite—curve 2, Bjurböle meteorite—curve 3, $\alpha = 10.95$ deg/min

As it is seen in Fig. 8, there are two well separated peaks for the rate of the argon release in biotite. The authors interpret this to be due to two differently bound components of argon.

Levskii concludes from the broad peaks for the rate of the helium release from several stony meteorites (Fig. 9), that a variety of similar activation energies is present. We think, it is possible that the broadening of the peaks is due to the various grain sizes. This fact may explain as well the slow high temperature tail of the curves resulting from a few large grains still having appreciable amounts of argon.

This is seen more clearly by the following theoretical discussion of the linear heating technique.

In case of a dependence of D on t (e.g. linear heating, $T = \alpha t$) from the equation $dc/dt = D \Delta c$ the corresponding solution of (4b) is derived by replacing Dt by the integral $I = \int D(t) \cdot dt$:

$$F = \frac{\lambda}{\sqrt{\alpha}} I^{1/2} - \frac{\mu}{\alpha} I; \quad \lambda = \frac{6}{R\sqrt{\pi}}, \quad \mu = \frac{3}{R^2} \cdot \quad (14)$$

The second derivative is found to be:

$$F'' = \frac{\lambda}{2\sqrt{\alpha}} \frac{D'}{I^{1/2}} - \frac{\lambda}{4\sqrt{\alpha}} \frac{D^2}{I^{3/2}} - \frac{\mu}{\alpha} D' .$$ (15)

From $D = D_0 \cdot \exp(-T_0/T)$ it follows:

$$D' = D \, T_0/T^2; \quad I = D \, T^2/T_0; \quad D'I = D^2 ;$$

and finally:

$$T_m \, e^{-\frac{T_0}{2T_m}} = \frac{R}{2} \sqrt{\frac{\alpha T_0}{\pi D_0}} .$$ (16)

The peak temperature, T_m, is related to the heating rate, α, and the activation temperature, T_0, as usual. In addition, however, T_m increases with the grain size. (This dependence does not exist for a chemical reaction in a system with impermeable walls.) Thus, if a mixture of grain sizes exists, a broadening will occur. The following expression is derived easily:

$$\Delta T_{m_{1,2}} = 4.6 \frac{T_{m_1} T_{m_2}}{T_0} \log \frac{R_1}{R_2} .$$ (18)

For a crude estimate we choose:

$$T_{m_1} \cdot T_{m_2} \sim T_m^2, \quad T_m \sim 700^0 \, \text{K} ,$$
$$T_0 = 25\,000^\circ \, \text{K.}$$

It follows:

$$\Delta T_{m_{1,2}} \sim 100^\circ \, \text{K} \quad \text{for} \quad R_1/R_2 = 10 ,$$
$$\Delta T_{m_{1,2}} \sim 200^\circ \, \text{K} \quad \text{for} \quad R_1/R_2 = 100 .$$

From two different heating rates the activation temperature T_0 is found, (for identical samples):

$$T_0 = \frac{T_{m_1} \cdot T_{m_2}}{T_{m_2} - T_{m_1}} \left\{ \ln \frac{\alpha_2}{\alpha_1} - 2 \ln \frac{T_{m_2}}{T_{m_1}} \right\} .$$ (17)

This equation is the same as given by GERLING et al. (1963), though their equation was derived from an equation describing first order chemical reactions.

5. Experimental results

5.1. Synthetic single crystals

The experimental conditions of the argon diffusion in natural samples are often far from ideal. Therefore the principles of diffusion are studied better in simple systems, where the diffusion geometry, the crystal structure, the impurity content etc. are well known. In addition it is very helpful, if some knowledge about the fundamental processes of the host lattice is present, such as the diffusion of the lattice ions.

This is true for the alkali halides, which are commercially available as single crystals of high purity grade and from which crystals of defined dimensions can be cleaved easily. By neutron irradiation, as described in 4.2, argon is introduced into the K-halides. The first investigations were performed with highly irradiated specimens, since the diffusion of Ar^{39} had to be measured down to low temperatures (Fechtig, Gentner,

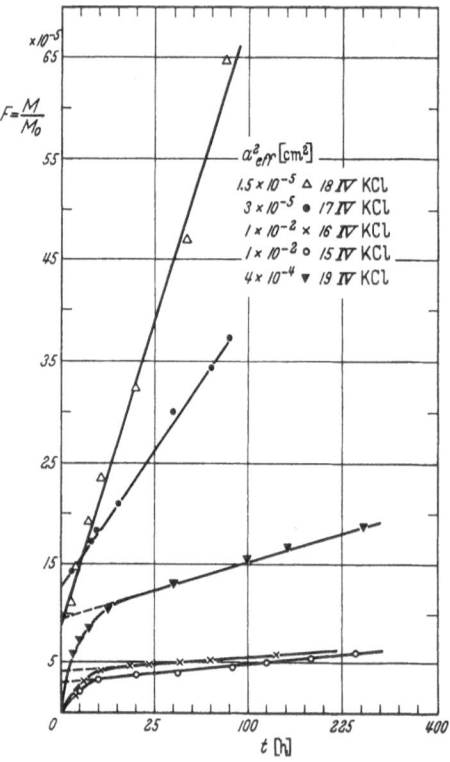

Fig. 10. Argon release from KCl at 107° C as a function of time. Powdered specimens, not annealed. The curved portion is due to a non-volumic diffusion process

and Kalbitzer, 1961; Kalbitzer, 1962). The number of argon atoms introduced was 10^{15} per cc. approximately. It was found that the argon release in the temperature range between 100 and approximately 600° C was characterized by a single activation energy for each K-halide. Corrections for nonideal gas release were made in the low temperature range.

At low temperatures, e.g. at 107° C (see Fig. 10), a fast diffusing fraction of argon, only several 10^{-5} parts of the total amount, was observed in KCl. Mechanical damage increased this irregular component, which is also frequently observed in irradiated natural samples. This exhaustible small amount characterized by very high diffusion constants was interpreted to be due to a non-volumic diffusion process along higher structural defects. If the irregular component was subtracted from the total amount released, the calculated diffusion constants agreed with the value extra-

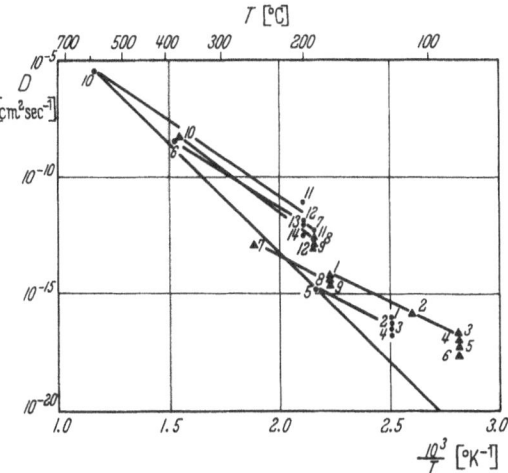

Fig. 11. Non-volumic diffusion of Ar in KCl, observed after temperature treatment; the sequence of the measurements is shown by the numbers. Cleaved crystals, not annealed

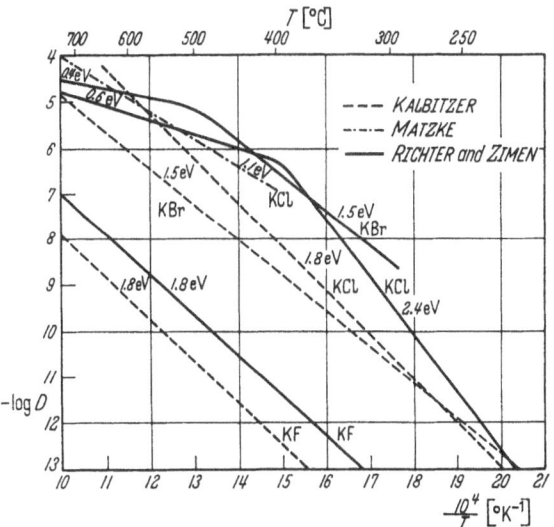

Fig. 12. Review of the results on volume diffusion of argon in KF, KCl, and KBr (after RICHTER and ZIMEN, 1965)

polated from the volume diffusion process at higher temperatures. When the sample was heated at higher temperatures and quickly cooled down, a similar effect was observed (Fig. 11).

The recent state of the research of argon diffusion in K-halides is shown in Fig. 12. There is only good agreement of the results for the system Ar/KF. Though for Ar/KBr the same activation energies are reported for the lower temperature range, the diffusion constants are in poor agreement. For Ar/KCl the corresponding data are discordant, too. In both systems the high temperature range deserves attention, as rather different values of the activation energies were found. Schmeling (1965) finds a pronounced effect of neutron dose on the diffusion of argon in KCl. This might indicate the influence of radiation damage. On the other hand a much lower dose means a considerable reduction of the number of argon atoms introduced. 10^{10} instead of 10^{15} argon impurities per cc might have quite a different fate during diffusion with respect to traps, dislocations and other lattice imperfections.

As to the interesting question of the diffusion mechanism all workers agree that the motion of argon is not correlated with the vacancy mechanism by which the K^+-ions migrate. In particular the high diffusion constants of argon at elevated temperatures being orders of magnitude higher than any known diffusion process of the lattice ions require a mechanism of its own. This also pertains to a part of the temperature region below 400° C. The interstitial mechanism could account for the high rates observed, although the two activation energies in the high temperature range observed by Richter and Zimen (1965) put up a difficult question.

These results on this "simple" system clearly show that the diffusion of argon at low temperatures should not be calculated from high temperature measurements, but that measurements have to be performed in the temperature interval of interest.

From the values of the diffusion constant for Ar in KCl measured at approximately 80° C, $D = 10^{-21 \pm 1}$ cm²sec⁻¹, it is seen that the salt mine of Buggingen at the initial temperature of approximately 80° C has suffered considerable fractional argon losses in the beginning (from the small absolute amount present at that time). We find a lowest loss parameter for grains of .3 mm radius of $d = 4\pi$, whereas the upper limit is $d = .4\pi$. (see Fig. 5b).

For the present temperature of 40° C the extrapolated diffusion constant is roughly 10^3 times smaller; thus the upper loss parameter is approximately 12π, which means small losses of argon by volume diffusion during the time of 20 m.y.

5.2. Minerals

The diffusion of gases in ideal and isotropic crystals, as discussed above for KCl, is more complicated for natural minerals. This was experimentally shown to be the case by Cook (1939), Flügge and Ziemens (1939), and others for the diffusion of argon from K-containing minerals.

This complication arises as the function

$$\log D \ vs \ \frac{1}{T} \quad or \quad \log D/a^2 \ vs \ \frac{1}{T}$$

is in general non-linear for the entire temperature range between 25° C and 1000° C. On the contrary the general result is several straight lines as shown in Fig. 13. Each of these straight lines has its own slope, which means a corresponding different activation energy E_1, E_2, E_3, ... Each of these activation energies E_i ($i = 1, 2, 3, ...$) represents a different diffusion mechanism dominant in a definite temperature range.

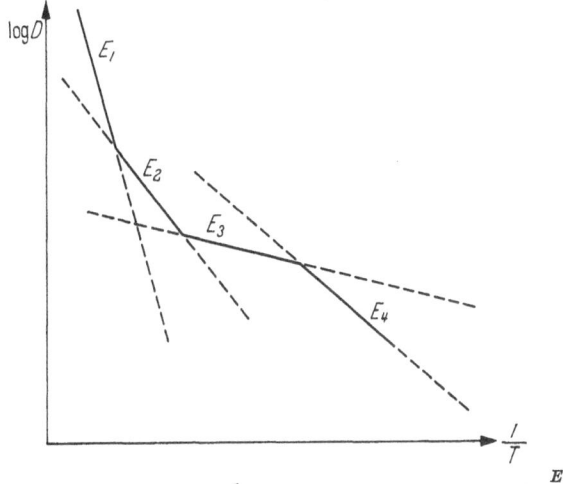

Fig. 13. Schematic curve $\log D$ vs $\frac{1}{T}$ given by the formula $D = D_0 \cdot e^{-\frac{E_1}{RT}}$

In principle one has for isotropic natural minerals at least two straight lines: one representing the volume diffusion for higher temperatures and the other representing non-volume diffusion for lower temperatures. The value of the activation energy of the volume diffusion is always considerably higher than the activation energy of the non-volume diffusion. In more complex lattices (micas) or in minerals with more than one component (perthite) there are sometimes more than two and occasionally even as many as five different values of the activation energies (GERLING and MOROZOVA, 1962).

Generally, it can be said that any change of the activation energy means a change in the diffusion mechanism of the neutral argon atoms in the lattice of the material. Such a change may be caused by any one of the following:

a) For nonisotropic lattices the localisation of the argon atoms at the beginning of a diffusion process may be different, energetically.

b) Phase transitions of the crystal lattice. Such transitions always cause a change of the lattice distances and thus a change in the value of the activation energy, in which case we always have volume diffusion.

c) Lattice imperfections which all natural crystals (minerals) show. For example inner surfaces, caused by low angle growing; line dislocations, caused by spiralic growth etc. Along such one- or two-dimensional dislocations the argon atoms may move easier than by volume diffusion.

d) Other effects, for example, absorption or desorption effects at inner surfaces may cause an apparent but no real change of the activation energy.

In the following discussion the experimental results will be interpreted with respect to these diffusion mechanisms. For possible age corrections the two following items are important:

1. If there are different reservoirs of argon for interesting minerals, are they or are they not dependent on each other;

2. How valid is the extrapolation of the diffusion constant D or the diffusion parameter D/a^2 obtained for higher temperatures down to interesting lower temperatures?

Gerling (1939) was the first to realize the practical meaning of the activation energy — as he called it in his paper the "diffusion heat" — as a criterion for age determinations of minerals. His first paper deals with the determination of activation energies for helium diffusion according to the Pb-He-method. This criterion is based on the fact that the gas loss of minerals at high temperatures (e.g. 1000° C) is of the same order of magnitude. For an extrapolation of the values of the diffusion constants from the experimental temperature range to room temperature, which is very often an extrapolation up to the 10th order of magnitude, only the slope of the extrapolated straight line is important. This is, therefore, only a question of the activation energy.

This criterion becomes questionable when the function $\log D$ vs. $1/T$ is not a single straight line. Unfortunately, this multiple slope situation is the case in many minerals to be dated. Gerling and Morozova (1957, 1958) and Gerling, Levskii, and Morozova (1963) have found spectra for the activation energy.

In the following sections the most important diffusion results will be reported.

There is a high abundance of potassium-containing minerals in the earth's crust. The potassium-argon method may be especially suitable for dating the micas and the feldspars. Data on coexisting micas and feldspars clearly show greater retentivity of argon for micas than for feldspars. Many papers and investigations have been published in this field: Aldrich and Wetherill (1958), Carr and Kulp (1957), Evernden and Richards (1962), Folinsbee, Lipson, and Reynolds (1956),

GOLDISH, BAADSGAARD, and NIER (1957), KAWANO and NEDA (1964), LIP-
SON, REYNOLDS, and FOLINSBEE (1956), WASSERBURG, HAYDEN, and
JENSEN (1956), WETHERILL, ALDRICH, and DAVIS (1955), WASSERBURG
and HAYDEN (1956) and others.

The comparison of the potassium-argon method to the Pb-Sr-method
or to coexisting uranium ages show that micas normally have a very
good retentivity for Ar^{40}. Therefore, the ages on micas are usually
concordant with other radioactive ages. Feldspars, on the other hand,
show larger age differences. Many papers have also been published on
this topic: ALDRICH (1956); ALDRICH and WETHERILL (1960); ALDRICH,
DAVIES, TILTON, WETHERILL, and JEFFREY (1956); ALDIRCH, WETHERILL,
DAVIES, and TILTON (1958); GAST, KULP, and LONG (1958); HURLEY,
CORNIER, HOWER, FAIRBAIRN, and PINSON JR. (1960a); KULP and
ENGELS (1963); LONG, COBB, and KULP (1959); TILTON, WETHERILL,
DAVIES, and HOPSON (1958); TILTON, WETHERILL, DAVIES, and BASS
(1960); WASSERBURG and HAYDEN (1955c, 1956); WETHERILL (1957);
WETHERILL, TILTON, DAVIES, and ALDRICH (1956b); and others.

5.2.1 Micas

Usually the direct diffusion experiments in the laboratory are carried
out with the aid of a mass spectrometer. Our own measurements were
carried out by the above mentioned radioactive method. Generally
— except the result of the most recent paper by BRANDT and VORONOVSKY
(1964a) — the function, $\log D$ $vs \frac{1}{T}$ or $\log \frac{D}{a^2}$ $vs \frac{1}{T}$, is non-linear for
micas. REYNOLDS (1957) and EVERNDEN et al. (1960) found for tem-
peratures greater than 250° C both lepidolithe and phlogopite are characterized by a curve which can be produced by a superposition of two
different straight lines. REYNOLDS (1957) has found that the argon
losses at higher temperatures (between 600 and 900° C) may be influenced by the expulsion of water. EVERNDEN et al. (1960) discussed
the function for diffusion parallel and perpendicular to the cleavage of
phlogopite. The steeper part represents the diffusion parallel to cleavage
whereas the flatter part represents the diffusion perpendicular to cleavage.
Furthermore, there is a remarkable influence of crystal water which
can change the dominant diffusion mechanism by influencing the lattice.
The authors extrapolated the flatter part of the function down to room
temperature with an activation energy of 28.0 kcal/mole and $D_{20°C} \approx$
$\approx 10^{-29}$ cm²/sec. Some diffusion results for glauconite show a linear
curve with $E = 28$ kcal/mole and $D_{20°C} \approx 10^{-29}$ cm²/sec which is similar
to that obtained for phlogopite. Glauconite turned out to be a very
suitable material for K-Ar-dating as extensively described in one of the
following chapters.

Experimental investigations carried out by Murina and Sprintsson (1961) and by Sardarov (1963) on the retentivity of radiogenic argon and the liberation of argon from glauconites led to the finding that argon is quantitatively preserved in the glauconites up to 400° C. Furthermore the liberation of this storedup argon, which takes place at 400° C, is connected with the expulsion of hydroxyl water. The conclusion was that the escape of argon is due to a structural decomposition of the mineral at this temperature. Sardarov calculated an activation energy of $E = 60$ kcal/mole which insures the complete retention of argon in glauconites at lower temperatures.

Gerling (1961), Gerling and Morozova (1957, 1962) and Gerling et al. (1963) have found that for muscovite the argon diffusion mechanism has three different activation energies which are determined to be 18, 37 and 72 kcal/mole or 33, 49 and 69 kcal/mole respectively. These authors have also determined activation energies for muscovite ($E = 85$ kcal/mole), biotite ($E = 57$ kcal/mole), and phlogopite ($E = 67$ kcal/mole). The conclusion, according to Gerling's criterion, is that these micas are suitable to be dated by the potassium-argon-method.

Contrary to the results of Gerling (1961) and Gerling and Morozova (1957, 1962), who discovered three different activation energies for muscovite, the most recent paper by Brandt and Voronovsky (1964a) shows, for temperatures higher than 300° C, only one single activation energy ($E = 55.8$ kcal/mole).

Own measurements (Fechtig, Gentner and Zähringer, 1960), using the radioactive method, were carried out for fluorite, anorthite, augite and margarite and the results observed clearly showed two and sometimes three different components (the third part for temperatures below 100° C). Thus, it is correct to assume the presence of two or three different activation energies respectively. The low temperature part (below 100° C) may be due to absorption of argon at inner surfaces. The middle part is not a volume diffusion and Kalbitzer (1962) hypothesized it may be due to a two- or even one-dimensional diffusion mechanism along lattice imperfections. Finally, the steep part for high temperatures (between 500 and 1000° C) represents the volume diffusion.

For materials, where the potassium is part of the chemical formula, nearly the entire amount of argon is lost by volume diffusion. Thus for margarite, the extrapolation of the volume diffusion down to room temperature is permissable. This results in a diffusion constant of $D_{20° C} < 10^{-30}$ cm²/sec and an activation energy of $E = 53.7$ kcal/mole.

A very interesting experiment was carried out by Hurley et al. (1962b). The authors tried to determine the diffusion parameter D/a^2 by determination of apparent ages on biotite. Since the samples were estimated to have originated at depths up to 9000 ft prior to a fault displacement, the mica was therefore exposed to temperatures in the neighborhood of 110° C. With the true age of the biotite compared with its apparent age it was possible to calculate $(D/a^2)_{110° C} = 6 \times 10^{-16} \sec^{-1}$. The authors give as a result the function D/a^2 vs $\frac{1}{T}$ with $D_0/a^2 = 1 \sec^{-1}$ and $E = 27$ kcal/mole.

Interesting investigations were also carried out by GERLING, MORO-ZOVA and KURBATOV (1961b) on the influence of the grinding process as a possible loss of argon. For this purpose micas were ground in a mechanical mortar, filed, and cut with scissors. For the cut muscovite $E = 93$ kcal/mole and 64% of argon are bound up with it. For two small mica fractions in the temperature range 500—800° C three values of activation energy were obtained, equalling 18, 37 and 72 kcal/mole. For the sample of ground mica argon losses were found to be 30% as shown in Fig. 14. Changes in the lattice distances by grinding were confirmed by X-ray diffraction.

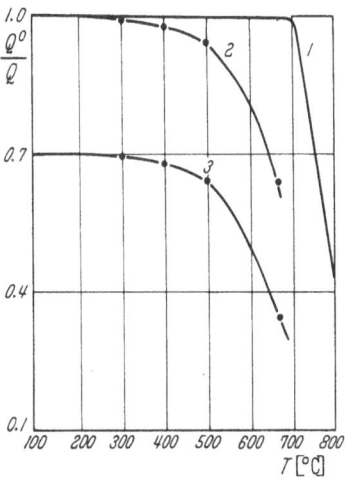

Fig. 14. Argon losses in mica for cut, filed and ground mica samples according to GERLING et al. (1961). Q = total content of argon; Q^0 = amount of argon remaining after heating for two hours; curve 1 = cut micas; curve 2 = filed micas; curve 3 = ground micas

Investigations by BRANDT and VORONOVSKY (1964b) on the influence of dehydration on the diffusion of radiogenic argon in micas led to the finding of three activation energies between 40 and 60 kcal/mole in connection with the dehydration.

KHUTSAIDZE (1962) investigated a possible influence of the air on argon losses from biotite in contrast to the same experiments in a vacuum. The author found that the loss of argon during heating of biotite depends upon the amount of oxidation of FeO that takes place. For a coarser fraction more oxidation occurs during heating (to 900° C) in air than in vaccum. A finer sized fraction heated in a vaccum loses more argon than a coarse fraction. The finer sized fraction also loses more argon in vacuum than in air. These results indicate that only fresh biotite can be used for geochronological purposes.

In a recent publication by FRECHEN and LIPPOLT (1965) new results of age determination for some samples of biotites were given, on which diffusion measurements were carried out. Activation energies of 69 and 86 kcal/mole and D/a^2 at 20° C of 10^{-29} and $10^{-34} \sec^{-1}$ for two different biotite samples were obtained.

5.2.2 Feldspars

Besides the micas, the potassium-bearing feldspars are suitable minerals to be dated by the potassium-argon method.

The first authors who carried out diffusion measurements on feldspars were NODDACK and ZEITLER (1956). They investigated the diffusion loss of argon for orthoclas samples. In 1957 GENTNER and KLEY carried out experiments on microcline, orthoclas and anorthoclas similar to those previously reported on the mica-ground experiments by GERLING (1961) for micas. Here feldspars were crushed down to the μ-range and a determination of argon for the several grain sizes was made.

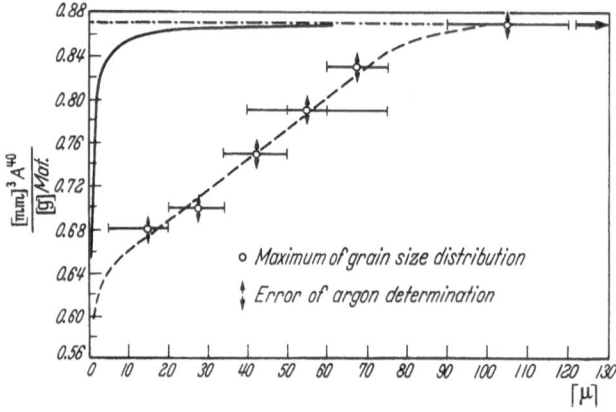

Fig. 15. Argon content in Microcline Varuträsk as a function of the grain sizes according to GENTNER and KLEY (1957). (------------ measured curve; ———— calculated curve)

The result is given in Fig. 15. The conclusion is as follows: due to metamorphic dislocations — since they are known especially for feldspars by processes like caolinization, perthitisation, serizitation, and emolinisation — feldspars produce inner surfaces and even holes or dislocations, in which argon may be collected. During the crushing process these areas preferentially break apart because they are energetically weaker than the nonviolated parts of the mineral. The smaller the pieces are, the more such inner surfaces must be broken off and the higher should be the argon loss.

Similar results have been obtained by STEVENS and SHILLIBEER (1956) in the case of perthites.

GERLING and MOROZOVA (1962) have found for microcline that the argon losses occur within five different ranges: with $E = 15, 26, 42, 99,$ and 130 kcal/mole. GERLING, MOROZOVA and KURBATOV (1961 b) have confirmed these five different values to be: $E = 15, 26, 42, 99,$ and 133 kcal/mole. These authors got an interesting result by grinding the microcline and carrying out the same experiment with the result that they have found only four different values for $E = 7.5$, 15, 24, and 115 kcal/mole. During this grinding process the amount of easily isolated argon increased from 20 to 70%. This is accompanied by a change of the activation energy values. Structural changes due to grinding are also revealed and are confirmed by X-ray diffraction.

GERLING et al. (1963) have also found for microline three different activation energies equalling 32, 46, and 100 kcal/mole.

Diffusion experiments carried out by REYNOLDS (1957) on feldspar show a bent curve for D vs $\frac{1}{T}$. This means that there are two different activation energies. The author concluded that it was allowable to extrapolate this curve resulting in a D value for $300°$ K of $D_{300°K} \approx 10^{-19}$ cm²/sec.

The result reported by EVERNDEN et al. (1960) shows a linear curve for microcline between 150 and 400° C. For higher temperatures there are some irregularities due to the change of a two-component feldspar into a one-component feldspar occurring between 400 and 800 °C. By extrapolating down to room temperature a diffusion constant of $D_{300°K} \approx 10^{-26}$ cm²/sec with $E \approx 24$ kcal/mole resulted.

In comparison to our own result (FECHTIG, GENTNER, and ZÄHRINGER, 1960) for anorthite, as described above, a differentiation must be made between a volume diffusion component and a non-volume diffusion component. Neglecting the non-volume component, we get a diffusion constant of $D < 10^{-30}$ cm²/sec at room temperature with an activation energy of 55.9 kcal/mole.

The diffusion loss for two feldspars investigated by AMIRKHANOV, BRANDT, and BARTNISKII (1959c) shows that argon is present in three different locations with $E = 32{-}35$, 93, and 106 kcal/mole. The authors extrapolated the steep part down to room temperature with the result of $D_{25°C} = 1.3 \times 10^{-72}$ cm²/sec. For the extrapolation of the flat part of the curve the result is $D_{25°C} = 7.4 \times 10^{-28}$ cm²/sec.

BAADSGAARD, LIPSON, and FOLINSBEE (1961) published a paper concerning the diffusion of argon in microcline and sanidine. The results are given in Fig. 16, which shows the function D vs $\frac{1}{T}$ with the following result:

1. microcline:

$E = 18$ kcal/mole $D_{25°C} = 10^{-20}$ cm²/sec.

2. sanidine:

$E = 52$ kcal/mole $D_{25°C} = 10^{-38}$ cm²/sec.

With the radioactive method (FECHTIG, GENTNER, and KALBITZER, 1961) sanidine yielded a straight line for $\log D/a^2$ vs $1/T$ (Fig. 17). This result was retained through temperatures between 200 and 1000° C and for several runs of the same sample. This resulted in a diffusion parameter $(D/a^2)_{20°C} = 2 \cdot 10^{-28}$ sec⁻¹ and an activation energy

$E = 39.8$ kcal/mole. This sample of sanidine, however, did not show any non-volume diffusion which means that the influence of the non-

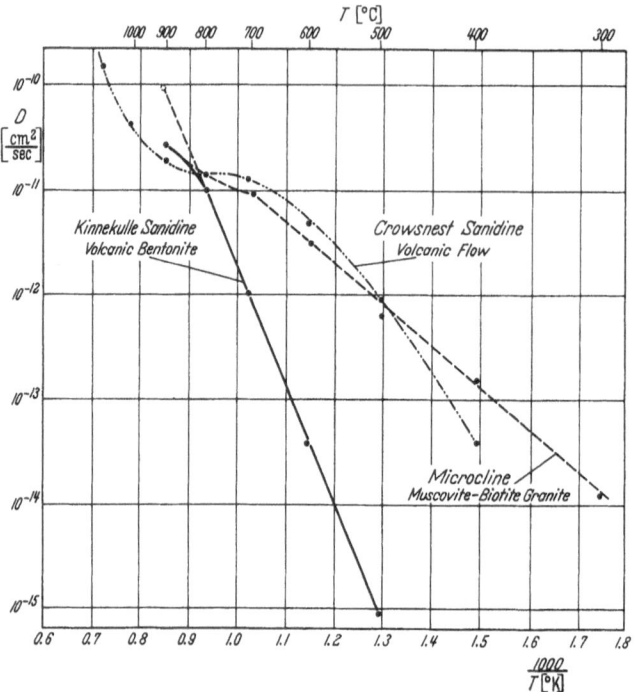

Fig. 16. The logarithm of the diffusion coefficient as a function of the reciprocal absolute temperature. (According to Baadsgaard et al., 1961)

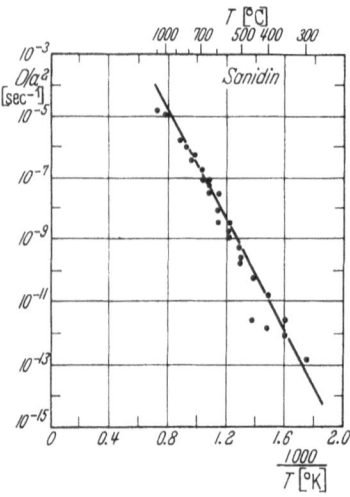

Fig. 17. D/a^2 as a function of $1/T$ for sanidine

volume diffusion for this sanidine is comparatively small to the other samples investigated with the radioactive method.

New results of age determination on two different sanidines have been published by FRECHEN and LIPPOLT (1965). These same samples were used for diffusion measurements and the results obtained were: activation energies of $E = 41$ and 48 kcal/mole and diffusion parameters D/a^2 ($20°$ C) $= 4 \cdot 10^{-28}$ and $3 \cdot 10^{-33}$ sec^{-1} respectively.

We conclude, therefore, that sanidine is the most suitable feldspar for use in the potassium-argon dating method. LIPPOLT (1961) successfully carried out datings on very young sanidine samples from the Eifel (Germany).

5.2.3 Other minerals

Besides these results on micas and feldspars, other minerals were investigated. For example, AMIRKHANOV, BARTNISKII, BRANDT and VOITKEVICH (1959a) have investigated the diffusion of argon in pyroxene with the result $E = 73.5$ kcal/mole and $D_{25°C} = 4 \cdot 10^{-55}$ cm^2/sec.

EVERNDEN et al. (1960) have investigated the diffusion of argon in leucite. They obtained an activation energy of $E = 44.2$ kcal/mole and $D_{0°C} = 10^{-33}$ cm^2/sec, which indicates leucite as a suitable material for dating with the K-Ar-method. Leucite shows a transition from a pseudoisometric structure to a true isometric habit. This transition takes place over a temperature range between 300 and 600° C. Due to this transition argon can be completely extracted at 550° C in one or two days as the authors showed in their investigations.

We have done some diffusion studies on other minerals (FECHTIG et al., 1960, 1961). Table 1 gives a survey of our data obtained with the radioactive method. The activation energy is that found for the volume diffusion. The diffusion parameter, D/a^2, is for all minerals appreciably smaller than 10^{-20} sec^{-1} and the correction parameter, $d^2 = \lambda a^2/D$, is always larger than $1000 \cdot \pi^2$. Non-volume diffusion, on the other hand, has much higher values for the diffusion parameter D/a^2 (up to 10^{-19} sec^{-1}). High argon losses may occur by this kind of diffusion mechanism.

HART (1961) compared the results of the K-Ar ages of hornblende samples and of pyroxenes with K-Ar ages of associated biotites, feldspars and zircons. He has found a good agreement of those ages and concluded that both hornblendes and pyroxenes are suitable for the K-Ar dating method.

5.3. Polycrystalline systems

Previous comparison between ideal single crystals and natural crystals (minerals) shows that the latter has more complicated the facts. Complexer polycrystalline systems show even further complications.

Table 1. *Activation energies, diffusion constants, diffusion parameters and age correction parameters for volume and non-volume diffusion*

Material	E (kcal/mol)	a (cm)	Volume diffusion			Non volume diffusion		
			$D_{20°C}$ (cm²/sec)	$(D/a^2)_{20°C}$ (sec⁻¹)	$t^2 = \lambda \cdot a^2/D$	$D_{20°C}$ (cm²/sec)	$(D/a^2)_{20°C}$ (sec⁻¹)	$t^2 = \lambda \cdot a^2/D$
Moldavit	30	$4 \cdot 10^{-2}$	$8 \cdot 10^{-27}$	$5 \cdot 10^{-24}$	$> 1000 \times \pi^2$	$< 8 \cdot 10^{-27}$	$< 5 \cdot 10^{-24}$	$> 1000 \cdot \pi^2$
Sanidin	40	$4 \cdot 10^{-2}$	$3 \cdot 10^{-31}$	$2 \cdot 10^{-28}$		$< 3 \cdot 10^{-31}$	$< 2 \cdot 10^{-28}$	$> 1000 \cdot \pi^2$
Phonolith	92	$< 5 \cdot 10^{-2}$	$< 1 \cdot 10^{-30}$	$< 1 \cdot 10^{-30}$		$2 \cdot 10^{-22}$	$1 \cdot 10^{-19}$	$17 \cdot \pi^2$
Augit	83	$1 \cdot 10^{-1}$	$< 1 \cdot 10^{-30}$	$< 1 \cdot 10^{-30}$		$1 \cdot 10^{-23}$	$1 \cdot 10^{-21}$	$1700 \cdot \pi^2$
Flusspat	44	$4 \cdot 10^{-2}$	$< 1 \cdot 10^{-30}$	$< 1 \cdot 10^{-30}$		$5 \cdot 10^{-23}$	$3 \cdot 10^{-20}$	$200 \cdot \pi^2$
Anorthit	56	$4 \cdot 10^{-2}$	$< 1 \cdot 10^{-30}$	$< 1 \cdot 10^{-30}$		$2 \cdot 10^{-23}$	$1 \cdot 10^{-20}$	$50 \cdot \pi^2$
Margarit	54	$5 \cdot 10^{-2}$	$< 1 \cdot 10^{-30}$	$< 1 \cdot 10^{-30}$		$3 \cdot 10^{-25}$	$1 \cdot 10^{-22}$	$> 1000 \cdot \pi^2$

A polycrystalline system is one composed of different minerals. Each mineral consists of different individual grain sizes. Meteorites are representatives of such polycrystalline systems.

The most correct but also the most complicated way to carry out diffusion studies on such polycrystalline systems would be to separate the different chemical components of the meteorites from one another and to carry out the experiments on these separated components. This procedure is usually applicable to rocks. Such separations were carried out mainly for other measurements by Merrihue (1964), Schaeffer et al. (1965) and Auer et al. (1965).

Unfortunately, in most cases we do not have enough meteoritic material to perform such separations. However, there is an alternative method to carry out diffusion studies on meteorites in spite of the above mentioned difficulty. The situation can be simplified by the assumption that the meteoritic components do not vary from each other in their diffusion behavior and that the determining factor is the grain size distribution of the meteorites. If we assume, furthermore, that the individual crystals are spheres with radius a, then we have to be concerned with a $\frac{1}{a^2}$ dependence of the diffusion loss. These simplifications are not correct but are good approximations especially if we keep in mind that we always need the diffusion parameters, D/a_{eff}^2, and not D or a_{eff} alone. The effective grain size, a_{eff}, may not be correct due to the above approximation.

Grain size distributions of meteorites have been published by Fechtig, Gentner, Lämmerzahl (1963) (Fig. 18). There was always a large number of fine grains

(Diameter $< 10 \ \mu$) which could not be counted. Extrapolation is, therefore, necessary for grain sizes below 10μ. According to the general dependence of $\frac{1}{a^2}$ this fine grained material might cause additional phenomena. The iron meteorite, Sikhote Alin, also showed different grain sizes. In kamazite grains with few mm diameters there

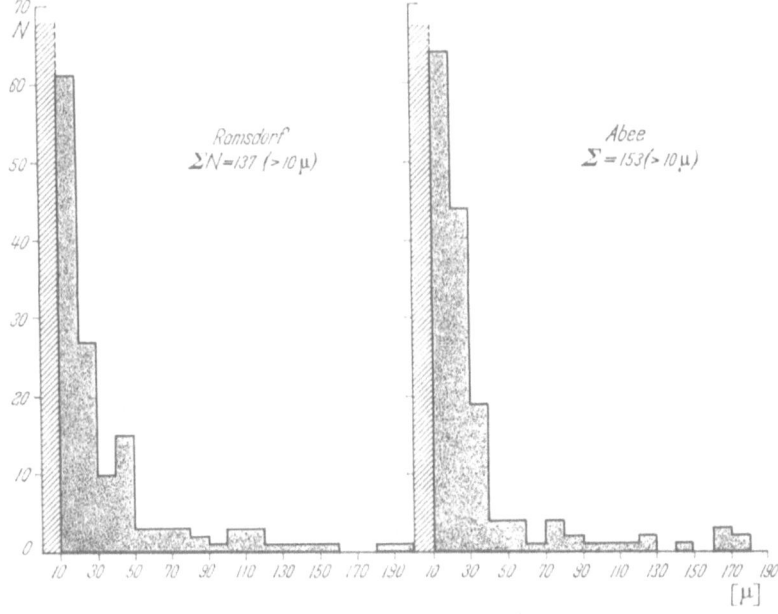

Fig. 18. Grain size distribution of some meteorites

are very small (few μ) needle-like parallel crystals consisting of taenite. Between the kamazite grains there are irregularly formed intrusions of schreibersite. This was also found by YAVNEL (1948).

Diffusion measurements were carried out for 9 different meteorites with varying K-Ar ages (see Table 2). Both the mass spectrometric method

Table 2. *List of the meteorites investigated with classifications, data of fall, and K-Ar ages*

Meteorite	Type	Fall	K–Ar-age
Alfianello	Hypersthene-Ch.	1883	0.77×10^9 y
Ramsdorf	Hyperstehne-Ch.	1958	0.42
Abee	Enstatite-Ch.	1952	4.71
Kapoeta	Howardite	1942	4.40
Tatahouine	Diogenite	1931	2.58
Nadiabondi	Bronzite-Ch.	1956	4.56
Bjurböle	Hypersthene-Ch.	1899	4.34
Bruderheim	Hypersthene-Ch.	1960	1.85
Sikhote-Alin	Fe-Hexaedrite	1947	—

or/and the radioactive method were applied. To use the radioactive method samples of meteorites were irradiated both with neutrons and with protons. In the first case Ar³⁷ is produced by the reaction Ca⁴⁰ (n, α) Ar³⁷. In the second case Ar³⁷ is produced by spallation processes of 600 Mev protons mainly with iron nuclei. All the irradiated samples were measured

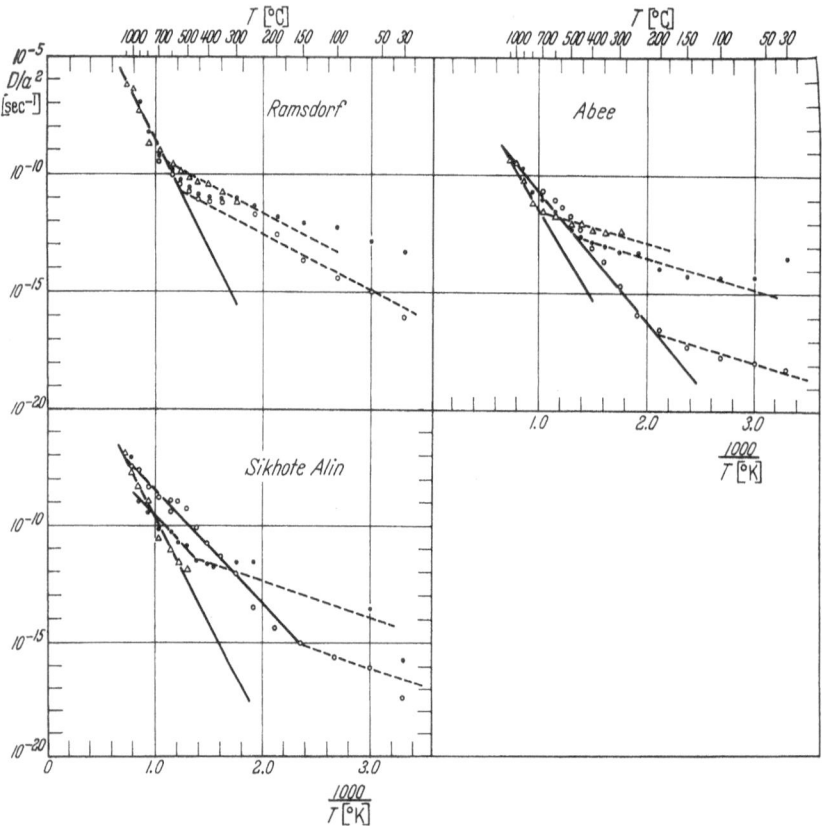

Fig. 19. Diffusion of Ar³⁷ in stone meteorites and in the iron meteorite Sikhote Alin
o first   ● second   △ third run

between 20° and 1000° C. According to equation (7) given by Wrage (1962) the diffusion parameter D/a^2 was determined and plotted as a function of $\frac{1}{T}$. The results are given for the proton irradiated samples in Fig. 19.

Correlating the results to the results in case of the minerals, two different straight lines can be seen: a steeper one for the higher temperatures and a flatter one for lower temperatures. Similarly, the steeper part represents the volume diffusion whereas the flatter part is representative of non-volume diffusion. Unlike previous results for minerals, the volume diffusion part is no longer reproducible for several runs. Instead, it always becomes steeper. This phenomenon can be explained as follows: with increasing degassing a_{eff} increases because the very small grains will be empty much earlier than the larger ones due to the $\frac{1}{a^2}$ dependence. This so-called "grain size effect" causes a change in slope. This explanation can be better understood by studying the mass spectrometric results.

Assuming that practically the entire argon escape is due to volume diffusion, an extrapolation of the steeper part to lower temperatures

Table 3. *Activation energies E and diffusion parameter D/a² at room temperature*

Meteorite	Ar³⁷ (irradiated with neutrons) E D/a^2		Ar³⁷ (irradiated with protons) E D/a^2		Ar⁴⁰ E D/a^2		Ar³⁸ E D/a^2	
Alfianello	(23)	(10⁻¹⁷)			45 (31)	10⁻²⁵ (10⁻²⁰)	(32)	(10⁻²³)
Ramsdorf	51	10⁻³⁶	43	10⁻³²	(27)	(10⁻¹⁷)	(27)	(10⁻¹⁹)
Abee	(30)	(10⁻²⁶)	34 (25)	10⁻²⁹ (10⁻²³)				
Kapoeta	39 (23)	10⁻²⁵ (10⁻¹⁹)						
Tatahouine	46	10⁻³⁵						
Nadiabondi					45 (27)	10⁻²⁸ (10⁻²⁰)	50 (31)	10⁻³¹ (10⁻²³)
Bjurböle					48 (22)	10⁻²³ (10⁻¹⁷)		
Bruderheim					(29)	(10⁻¹⁸)		
Sikhote-Alin			58 (24)	10⁻³⁰ (10⁻²⁰)			62 (26)	10⁻³⁶ (10⁻²¹)

is appropriate. The values for D/a^2 at 30° C, as well as the activation energies for the volume diffusion, are given in Table 3.

The values in brackets belong to the first run, those without brackets to the last run. A comparison between Ramsdorf and Abee shows no

larger differences for the different kinds of irradiation. The differences
indicated in the table are due to the special localization of Ca in these
meteorites.

Some other experiments on meteoritic samples using the mass
spectrometric method were carried out. For a definite temperature a

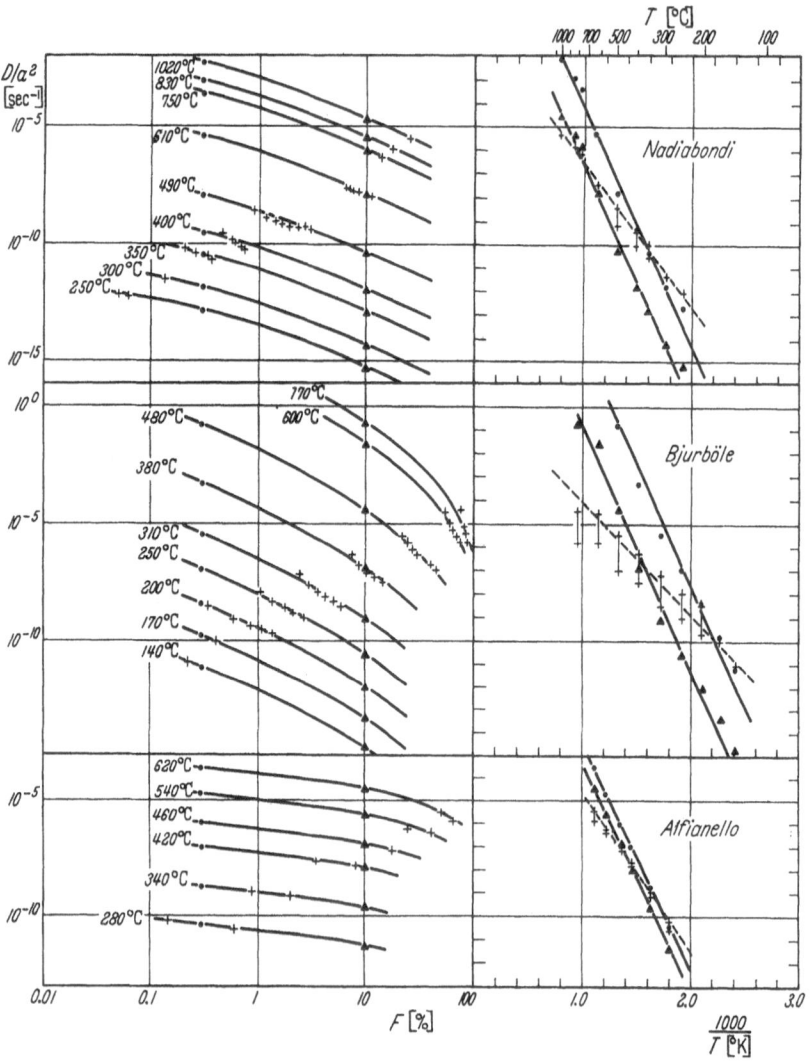

Fig. 20. Isothermal diffusion of Ar^{40} in stone meteorites. Left: grain size effect for isothermal degassing. Right: $\log D/a^2$ vs $1/T$ directly ---------- corrected for grain size effect ——— for F = 0.3
and 10%

fresh sample was used to study the argon loss as a function of time. This
procedure was carried out for a set of temperatures using fresh samples

for each temperature. As a result we found D/a^2 to be a function of the degassing rate. According to the relation

$$\frac{D}{a^2}(T, F) = D(T) \times \frac{1}{a^2(F)}$$

D/a^2 is represented in the $\log D/a^2 - \log F$ — diagram by a series of curves with T as a parameter in such a way that the distances in direction of the $\log D/a^2$ — scale are always the same for two different curves. The result is given in Fig. 20 where it can be seen that the experimental points are located on those curves. The right half of Fig. 20 shows the usual diagram $\log D/a^2$ vs $\frac{1}{T}$. Here both the values for D/a^2 corresponding to $F = 0.3$ and 10% (two solid straight lines) as well as the experimental values for D/a^2 (broken straightline) are plotted. Using this procedure the "grain size effect" was corrected. The effect of this correction is represented by a much higher activation energy.

All the results are listed in Table 3 both for the corrected and non-corrected values of E and D/a^2 at 25° C. It can further be seen that the radioactive and the mass spectrometric method give practically the same results. Our conclusions are therefore:

a) Assuming radioactive argon can escape only by volume diffusion and the meteorite to be dated was never exposed to temperatures higher than 30° C, no incorrect ages can be caused by diffusion.

b) Alfianello which shows only a weak correction of the grain size effect obviously has suffered an appreciable loss of argon. It cannot be decided whether the meteorite was exposed to higher temperatures or has been subject to shocks by hypervelocity impacts. This fact agrees with the very low K-Ar age.

Similar investigations were carried out by RUTSCH (1962) and BIERI and RUTSCH (1965) for samples of the meteorites Bruderheim and Holbrook. In Table 4 the activation energies for the diffusion of Ar[38] and Ar[40]

Table 4. *Activation energy E and diffusion parameter D/a^2 at 0° C for Bruderheim and Holbrook according to* BIERI *and* RUTSCH (1965)

Meteorite	E kcal/mole		D/a² (sec⁻¹)	
	Ar[38]	Ar[40]	Ar[38]	Ar[40]
Bruderheim	26	34	1×10^{-21}	5×10^{-24}
Holbrook	—	28	—	2×10^{-19}

found by the above mentioned authors are given. Although they discussed the grain size effect, the values in Table 4 are appreciably lower than the values we have found. The activation energy for Ar[40] corresponds to our own non-corrected value of $E = 29$ kcal/mole. Grain size effect on the value of the activation energy is extremely sensitive. In addition it occurs even at very low degassing rates. We agree with the conclusions concerning possible argon loss by diffusion.

5.4. Glasses

A short part of this article may be devoted to the diffusion results in amorphic materials, namely the glasses. Age determinations, according to the potassium-argon method, were carried out recently on tektites and impact glasses. Using the radioactive method we have investigated

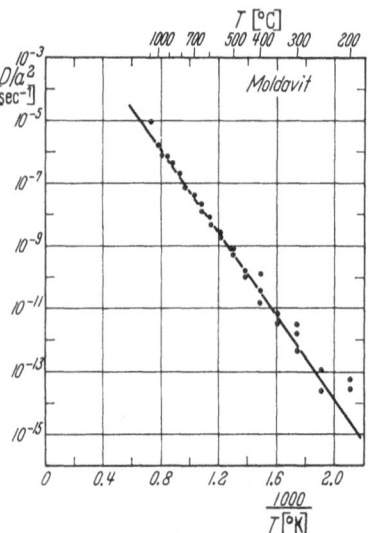

the diffusion behavior of argon in a tektite (moldavite). (FECHTIG, GENTNER and KALBITZER, 1961). The result is presented in Fig. 21. Several runs for the same irradiated sample were carried out. The result in the whole temperature range below 1000° C was a reproducible straight line. The activation energy was calculated to be $E = 29.5$ kcal/mole. The extrapolated value for D/a^2 at 20° C is $(D/a^2)_{20° C} = 5 \times 10^{-24}$ sec^{-1}. Using this value, any diffusion loss is impossible as long as the storage temperature for the material is not appreciably higher than room temperature.

Fig. 21. log D/a^2 for tektite glass (moldavite) as a function of $1/T$

This result is important for the reliability of the potassium-argon ages of the tektites. It should be mentioned that we found no non-volume components in these investigations.

NORTON (1953, 1957, 1961, 1962) has published some very interesting papers concerning the diffusion of gases through glasses. The investigations were performed as follows: The gas was sealed in a glass container. As a function of temperature the amount of gas which escapes through the glass by diffusion was measured. According to NORTON (1962) the permeation rate, P, was measured for argon through vitreous silica glass. Through the relation

$$P = D \times S \text{ with } D = \text{diffusion constant}$$

$$S = \text{solubility}$$

it was possible to calculate the diffusion constant D. Comparable to our values in the case of tektite material is the activation energy. NORTON found for vitreous silica glass $E = 32.7$ kcal/mole; our value for moldavite is 29.5 kcal/mole. These two values are comparatively close together, taking into account the differences between the several kinds of glasses used.

NORTON also found for various gases through several glasses the function P vs $\frac{1}{T}$ $\left(\text{respectively } D \text{ vs } \frac{1}{T}\right)$ tobe a single straight line for the whole temperature range. This leads to the conclusion that only one activation energy was found and therefore only one single diffusion process.

5.5. Conclusions

Independent studies reveal that (at least in many minerals) several reservoirs of radiogenic argon exist all energetically different from each other. The most reasonable extrapolation for the curve $\log D$ vs $1/T$ resp. $\log D/a^2 =$ vs $1/T$ down to interesting temperatures is dependant upon the amount of radiogenic argon located in these different reservoirs and upon a possible dependence of the reservoirs on each other. The results of GERLING et al. (1961, 1963) and AMIRKHANOV et al. (1959 a, c) given for micas and feldspars do not show any specific preference for some of the reservoirs. The results of the radioactive method in the case of KCl (KALBITZER, 1962) show that the amount of argon present in the lattice dislocations is far below 1% of the total amount of argon. KALBITZER (1962) also pointed out that these two reservoirs are connected.

Although the situation is quite complex one can say that at least for minerals which have K homogeneously distributed throughout the mineral practically all the argon can escape only by volume diffusion, if we do not take into account any metamorphism. For such minerals it is, therefore, allowed to extrapolate the straightline which represents the volume diffusion down to the temperatures investigators are most interested in. This section concludes that diffusion at room temperature is always so small that no appreciable argon losses occur. Obvious argon losses, especially for feldspars, are not due to diffusion effects but due to metamorphic structural dislocations.

6. The thermal history of minerals

The calculation of a mineral's temperature history is very often an interesting and practical endeavor. This can be done if the real age and diffusion parameters $(E, D/a^2)$ of a sample are known and the mineral has suffered diffusion losses.

This application is possible because:

a) For minerals the activation energies usually exceed 30 kcal/mole. This means that the term, D/a^2, is very sensitive to changes of the temperature.

b) At low temperatures (~ 20° C) this sensitivity is very great because of the $1/T$ dependence. (On the $1/T$-axis the distance between 800° C and 200° C is very similar to the distance between 200° C and room temperature.)

Such an application was made for the dating of the sylvite deposit in Buggingen by Smits and Gentner (1950) and Gentner, Präg and

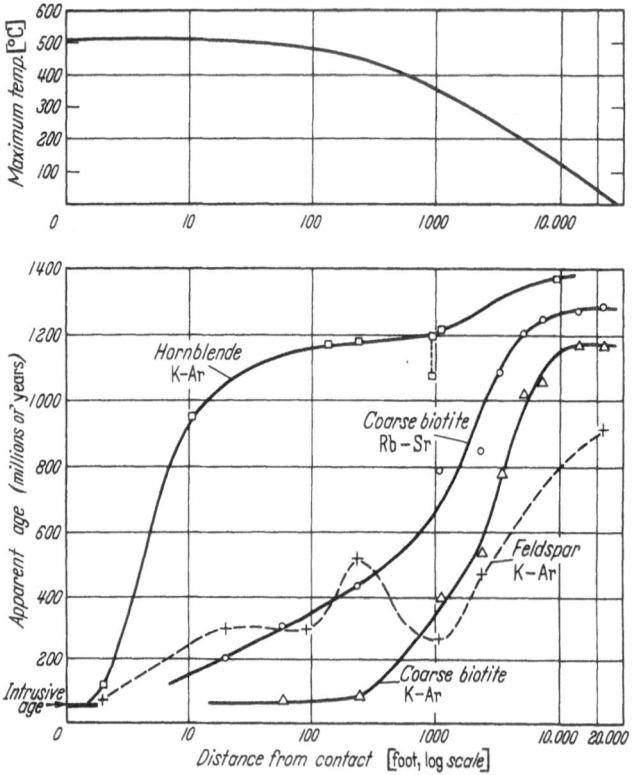

Fig. 22. Variation of apparent ages of minerals and of the theoretical maximum temperatures as a function of distance from an intrusive contact according to Tilton and Hart (1963)

Smits (1953 a, b). The authors had to consider a temperature history of the salt deposit in order to explain the diffusion losses they had measured. It had to be assumed that the salt deposit was initially at 80° C dropping to the present 40° C. Such a change indicates a variation of D/a^2 for five orders of magnitude.

Goles, Fish and Anders (1960) tried to calculate the temperature to which chondrites are exposed in space. The idea was to interpret the difference between the K-Ar ages and the Rb-Sr ages of chondrites as to be due to diffusion losses of radiogenic argon at that temperature.

To determine a straight line, D/a^2 vs $1/T$, the authors used degassing data given by GEISS and HESS (1958).

This straight line was then extrapolated over a wide temperature range. Because the authors could not take into account the grain size effect, the activation energy they used was 14 kcal/mole. This value is not in agreement with later data.

The most interesting application so far was given by TILTON and HART (1963) and HART (1964).

For precambrian rocks located near tertiary intrusive stock he determined the apparent K-Ar and Rb-Sr ages for hornblende, boitite, and feldspar (Fig. 22). These ages are plotted as a function of the distance from the intrusive stock. Taking into account some plausible assumptions HART succeeded in the following interpretation: The ages were produced by diffusion losses of Ar and Sr in these minerals. Assuming 500° C for the initial temperature of the intrusive stock with a true age of 54 m.y., and using a heat flow model, HART concluded that the surrounding material was warmed up thus causing diffusion losses. He could interpret the age distribution in biotites: The activation energies for argon and strontium in biotite are the same ($E = 30$ kcal/mole) and the difference in their diffusion coefficients is about a factor of five. The differences between the K-Ar and Rb-Sr ages are, therefore, caused solely by this factor of 5 for the diffusion constants.

7. Other possible influences on the K-Ar ages

As we have seen, not all potassium-bearing materials are equally suitable for dating according to the potassium-argon method. It was pointed out that the most suitable materials are the micas, sanidines and glasses.

Incorrect dating can result from argon losses. There are, however, other possible sources of error. As we have seen in the case of feldspars, the metamorphic structural change is a serious source of argon losses. SARDAROV (1957b) measured the argon loss in microcline as a function of an increasing effect of perthitization. He has shown quantitatively

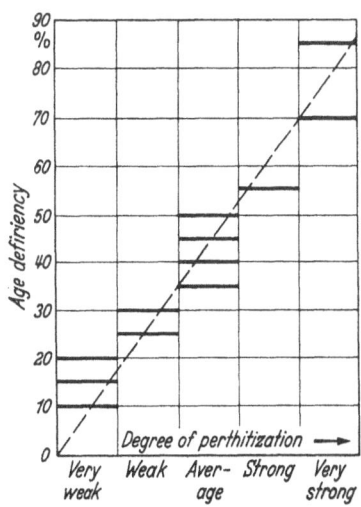

Fig. 23. Age deficiency of microcline in percent vs the degree of its perthitization according to SARDAROV (1957b)

that the apparent potassium-argon age decreases with increasing perthitization. This experimental result is represented in the plot of Fig. 23.

Table 5. *Diffusion constants D or diffusion parameters D/a^2 at room temperature and activation energies E for minerals, meteorites and glasses*

Material	Author	D, D/a^2 (Volume Diff.)	E kcal/mole	Method
Micas:				
biotite	GERLING and MOROZOVA (1957)	$(D/a^2)_{20°C} = 10^{-20} \text{sec}^{-1}$	57	direct, ms
biotite	HURLEY et al. (1962 b)		27	indirect
biotite	GERLING et al. (1963)	$(D/a^2)_{20°C} = 10^{-29} \text{sec}^{-1}$	69, 48, 33	direct, ms
biotite	FRECHEN and LIPPOLT (1965)	$(D/a^2)_{20°C} = 10^{-34} \text{sec}^{-1}$	69	direct, ms
biotite	FRECHEN and LIPPOLT (1965)		86	direct, ms
muscovite	GERLING et al. (1961 b)		72, 37, 18	direct, ms
	GERLING and MOROZOVA (1961)			
muscovite	GERLING and MOROZOVA (1957)		85	direct, ms
muscovite	BRANDT and VORONOVSKY (1964a)		56	direct, ms
phlogopite	EVERNDEN et al. (1960)	$D_{20°C} = 10^{-29} \text{cm}^2\text{sec}^{-1}$	28	direct, ms
phlogopite	GERLING and MOROZOVA (1957)		67	direct, ms
glauconite	EVERNDEN et al. (1960)	$D_{20°C} = 10^{-29} \text{cm}^2\text{sec}^{-1}$	28	direct, ms
glauconite	SARDAROV (1963)		60	indirect
margarite	FECHTIG et al. (1960)	$D_{20°C} = 10^{-30} \text{cm}^2\text{sec}^{-1}$	54	direct, rad.
Feldspars:				
feldspar	AMIRKHANOV et al. (1959 c)	$D_{25°C} = 10^{-72}, 10^{-27} \text{cm}^2\text{sec}^{-1}$	106, 93, 33	direct, ms
feldspar	REYNOLDS (1957)	$D_{25°C} = 10^{-19} \text{cm}^2\text{sec}^{-1}$		direct, ms
microcline	GERLING and MOROZOVA (1962)		130, 99, 42, 26, 15	direct, ms
microcline	GERLING et al. (1963)		100, 46, 32	direct, ms
microcline	EVERNDEN et al. (1960)	$D_{20°C} = 10^{-26} \text{cm}^2\text{sec}^{-1}$	24	direct, ms
microcline	BAADSGAARD et al. (1961)	$D_{25°C} = 10^{-20} \text{cm}^2\text{sec}^{-1}$	18	direct, ms
anorthite	FECHTIG et al. (1960)	$D_{20°C} = 10^{-30} \text{cm}^2\text{sec}^{-1}$	56	direct, rad.
sanidine	FECHTIG et al. (1961)	$(D/a^2)_{20°C} = 2 \cdot 10^{-28} \text{sec}^{-1}$	40	direct, rad.
sanidine	BAADSGAARD et al. (1961)	$D_{25°C} = 10^{-38} \text{cm}^2\text{sec}^{-1}$	52	direct, ms
sanidine	FRECHEN and LIPPOLT (1965)	$(D/a^2)_{20°C} = 4 \cdot 10^{-28} \text{sec}^{-1}$	41	direct, ms
sanidine	FRECHEN and LIPPOLT (1965)	$(D/a^2)_{20°C} = 3 \cdot 10^{-33} \text{sec}^{-1}$	48	direct, ms

Other minerals: sylvite (red)	AMIRKHANOV et al. (1959 b)	$D_{27°C} = 10^{-30}$ cm²sec⁻¹	37—47	direct, ms
sylvite (pink)	AMIRKHANOV et al. (1959)	$D_{20°C} = 10^{-30}$ cm²sec⁻¹	37—47, 16	direct, ms
sylvite synth.	KALBITZER (1962)	$= 10^{-26}$ cm²sec⁻¹	42	direct, rad.
pyroxene	AMIRKHANOV et al. (1959)	$D_{25°C} = 4 \cdot 10^{-55}$cm²sec⁻¹	74	direct, ms
leucite	EVERNDEN et al. (1960)	$D_{0°C} = 10^{-33}$cm²sec⁻¹	44	direct, ms
fluorspar	FECHTIG et al. (1960)	$(D/a^2)_{25°C} < 10^{-30}$sec⁻¹	44	direct, rad.
augite	FECHTIG et al. (1960)	$< 10^{-30}$sec⁻¹	83	direct, rad.
phonolithe	FECHTIG et al. (1960)	$< 10^{-10}$sec⁻¹	92	direct, rad.
Meteorites: Alfianello	FECHTIG et al. (1963)	$(D/a^2)_{25°C} = 10^{-25}$sec⁻¹	45	direct, ms
Ramsdorf	FECHTIG et al. (1963)	$= 10^{-36}$sec⁻¹	51	direct, rad.
Abee	FECHTIG et al. (1963)	$= 10^{-29}$sec⁻¹	34	direct, rad.
Kapoeta	FECHTIG et al. (1963)	$= 10^{-25}$sec⁻¹	39	direct, rad.
Tatahouine	FECHTIG et al. (1963)	$= 10^{-35}$sec⁻¹	46	direct, rad.
Nadiabondi	FECHTIG et al. (1963)	$= 10^{-28}$sec⁻¹	45	direct, rad.
Bjurböle	FECHTIG et al. (1963)	$= 10^{-23}$sec⁻¹	48	direct, ms
Sikhote Alin	FECHTIG et al. (1963)	$= 10^{-30}$sec⁻¹	58	direct, ms
Bruderheim	BIERI and RUTSCH (1962, 1965)	$(D/a^2)_{0°C} = 5 \cdot 10^{-24}$ sec⁻¹	34	direct, rad.
HOLBROOK	BIERI and RUTSCH (1962, 1965)	$= 2 \cdot 10^{-19}$ sec⁻¹	28	direct, ms
Glasses: tektite (moldavite)	FECHTIG et al. (1962)	$(D/a^2)_{20°C} = 5 \cdot 10^{-24}$sec⁻¹	30	direct, ms
vitreous silica glass	NORTON (1962)		33	direct, ms

Many papers deal with metamorphism and its influence on the results of the K-Ar ages of those minerals. Another chapter of this book represents the main results of those studies.

KULP and ENGELS (1963) have published interesting results of their experiments carried out on biotites. By the action of the cations (particularly calcium) in ordinary ground water, biotite is susceptible to base exchange whereby 50% of the potassium may be removed by this process. Fortunately, the influence on the K-Ar ages is always below 10% because argon is also removed by such a process.

Another possible influence on the K-Ar ages of minerals has not yet been investigated: The diffusion of potassium. At 20° C the diffusion of K is of the same order of magnitude as it is for the diffusion of Ar. A possible influence of the potassium diffusion cannot be excluded. Intensive investigations concerning the diffusion of Sr in connection with the Rb-Sr dating method have been carried out by McNUTT (1964). These studies show an appreciable influence of such a diffusion even in this case where the daughter product is a solid element.

For the K-Ar method the influence of potassium diffusion on dating results can be minimized by the choice of the material to be dated. If potassium is a main chemical constituent, the potassium has a homogenous distribution in the solid. Such may not be the case if K is present as a trace element.

8. Summary

An overall summary on the entire diffusion results of argon in K-bearing solids giving the investigated minerals, diffusion constants D or diffusion parameters D/a^2 respectively for room temperature, the activation energies, and the applied methods are drawn up in Table 5. The diffusion constants D or diffusion parameters D/a^2 are in most of the cases so low that any influence due to the diffusion of argon on the measured K-Ar age is negligible. We are however confronted with two exceptions in Table 5: 1. HURLEY et al. published $D/a^2 = 10^{-20}$ sec^{-1} for biotite at room temperature. This was an indirect method and the activation energy of $E = 27$ kcal/mole seems to be too low. 2. The value for feldspar given by REYNOLDS to be $D_{300° K} \approx 10^{-19}$ cm^2/sec, which is not an extrapolation of a straight line representing volume diffusion and might therefore be too high. All the other values are extremely low at room temperature. The situation, however, changes if one has to take into account higher temperatures, which is obvious in the case of sylvite where — as pointed out earlier — such an influence has to be taken into consideration.

Usually it is not practical to carry out age corrections because of the fact that a small variation in the temperature easily causes a variation of D/a^2 by several orders of magnitude.

In spite of the fact that there are still large differences in the results between various research groups, it should be emphasized that there are two main advantages and one main disadvantage of the radioactive method in comparison to the mass spectrometric technique:

a) By irradiation an Ar^{37} concentration up to 10^8 cps can be produced. A common gas counter with a sensitivity of a few cps therefore allows that the range for measurements is up to the 8th order of magnitude.

b) As pointed out earlier, it is very important for diffusion studies to know the initial gas concentration in crystals. By applying the radioactive method and by using suitable experimental conditions it is always possible to produce a homogenous gas concentration. This is of special interest for meteorites where results can be extremely influenced by non-homogenous initial gas concentrations.

c) The disadvantage is as follows: very often in nature — and especially for meteorites — K is present together with Ca. In this case both Ar^{37} and Ar^{39} concentrations will be produced by irradiation with similar cross sections (besides shorter lived radioactive Ar isotopes.) Due to the enormously different decay times (35 days and 325 years for Ar^{37} and Ar^{39} respectively) the measured activity is always the Ar^{37} activity, unless the time between the irradiation and the diffusion experiments is long enough so that the Ar^{37} activity is negligible. Differences arising between these results and results carried out with the Ar^{40} isotope may be due to different locations for Ca and K in non-homogenous solids.

The general results on the Ar diffusion and its influence on the K-Ar ages could be summed up as follows: The investigated material does not lose any argon over geological time distances by diffusion as long as the storage temperature does not appreciably exceed the room temperature. Serious influences, however, arise due to metamorphic events, for which some perthitizations in the case of feldspars may be an example. The diffusion behavior may be on the other hand a sensitive method for non-metamorphosed material to use as a thermometer.

K-Ar dating of Precambrian Rocks

By

G. W. WETHERILL

Introduction

Much of the early work in the development of the K-Ar dating method involved the measurement of radiogenic argon in Paleozoic and Precambrian rocks (ALDRICH and NIER, 1948; GERLING et al., 1949; MOUSUF, 1952; RUSSELL et al., 1953; WASSERBURG and HAYDEN, 1954b). One

reason for this was that the concentration of radiogenic argon in feldspar and mica from these rocks was considerably greater than that
obtainable from younger rocks and thus analytical difficulties were
minimized. This is illustrated in Fig. 1 which shows the concentration of
radiogenic argon as a function of time in one gram of potassium mineral
containing 10% K. Using an extraction system free of leaks but without
special precautions with regard to preliminary outgassing or removal
of atmospheric argon from the sample the amount of atmospheric
argon contamination is usually about 10^{-5} cc STP.

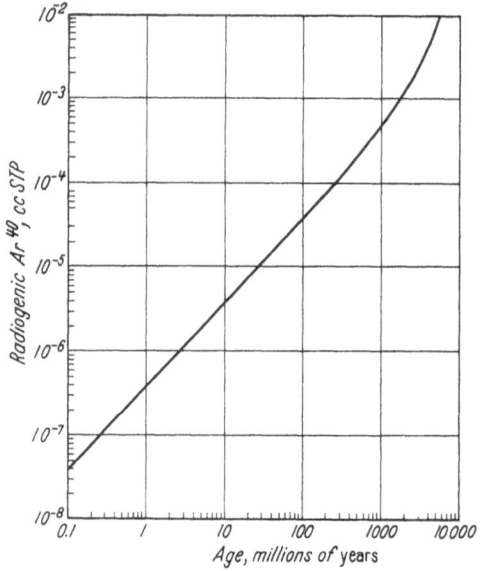

Fig. 1. Quantity of radiogenic argon which has been produced in one gram of a mineral containing
10% K (i.e., 100 mg of K) as a function of age. This quantity is proportional to the amount of K

From Fig. 1 it can be seen that a one gram sample of a 10% K
mineral analyzed under such conditions will be 45% radiogenic if it is
20 million years old, 89% radiogenic if it is 200 million years old and
99% radiogenic if it is 2000 million years old. All of this early work with
the exception of the single measurement by INGHRAM et al. (1950)
up to the time of the measurements of WASSERBURG and HAYDEN
(1954 b) involved volumetric measurement of the argon concentration
which could only be attempted with large ($\sim 10^{-3}$ cc STP or greater)
quantities of argon. Such quantities of argon could be readily obtained
from relatively small samples of Precambrian rocks, but required large
(~ 150 g) samples of even the very K-rich mineral sylvite when Tertiary
rocks were studied (SMITS and GENTNER, 1950). Another advantage of
working with relatively large samples was the greater ease of obtaining

large sample to background ratios in the mass spectrometer, thus mini-
mizing the error introduced by the uncertainty in the contribution of
the background peak at mass 36 to the correction for atmospheric argon.
Since in this early work there were a great many other problems, both
geological and analytical, to be discovered, evaluated, and overcome
it was in some ways desirable to work with the larger samples obtained
from the older rocks and thus reduce the number of difficulties which
had to be dealt with simultaneously.

There was, however, another compelling reason for working with
these older rocks, particularly with those of Precambrian age. This
was the fact that although more than 85% of earth history is Precambrian,
very little of the Precambrian geological record was well understood,
principally due to the absence of fossils which form the basis for long-
range correlation of Paleozoic and younger rocks. Thus the opportunity
for making major contributions to earth history was greatest for those
very rocks which were easiest to measure. For example, among the first
Precambrian ages to be measured on rocks which had not previously
been studied by U-Pb methods were K-Ar and Rb-Sr ages on a sample
of purple muscovite (originally misidentified as lepidolite) from the
Bridger Mountains, near Bonneville, Wyoming, U.S.A. Since there were
no previous age measurements on rocks in this entire part of the continent, there was no way to anticipate the result of these measurements.
The measured age (K-Ar = 2570 m.y., WETHERILL et al., 1954) has
subsequently been revised to 2250 m.y. (ALDRICH et al., 1958), as a
result of improved knowledge of the potassium decay constants. Furthermore, it is likely that the true age of this pegmatite is about 2600 m.y.
and that some radiogenic argon has been lost. Nevertheless, the exciting
and almost qualitative fact that there were early Precambrian rocks
in this region became known well before the later refinements in the
method.

Loss of radiogenic argon

It should not be overlooked that there were serious disadvantages
to this early attention with Precambrian ages, and that these difficulties are still present. These disadvantages are all fundamentally
related to the fact that in nature, rocks and minerals usually lose at
least some, and frequently much of the radiogenic argon generated
within them since they were formed. In some cases, these losses may
not be related to any otherwise identifiable geological cause and probably result from low temperature diffusion of argon from the crystalline
lattice to cracks and other imperfections in the mineral from which
the argon readily escapes. This cause is probably responsible for the
loss of ~ 20% of the radiogenic argon from microcline and perhaps a

greater amount from natural glasses. In addition, relatively low temperatures, i.e., 300—500° C maintained over million of years can result in
partial or complete loss of radiogenic argon from micas, especially biotite,
and with rather more difficulty, from amphiboles. In such circumstances,
other evidences for metamorphism are frequently, but not always, present.
These problems have been discussed in more detail in previous chapters.

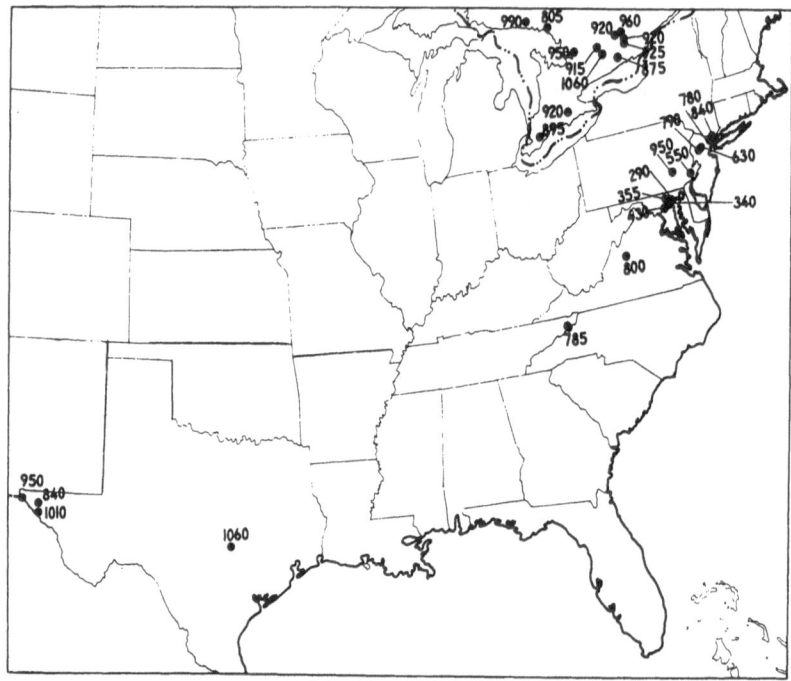

Fig. 2. K-Ar ages of minerals from rocks of Grenville (~ 1100 m.y.) age in part of N. America

As a consequence of this argon loss, K-Ar ages are frequently too
low. This problem is most serious for older rocks because those older
(e.g., Paleozoic or Precambrian) rocks now exposed at the surface almost
always have been at some great depth within the earth, perhaps many
kilometers, at which time they were at a higher temperature thus
facilitating the loss of argon. Also the fact that the rocks are older increases the probability that later extensive periods of metamorphism
and magmatism will have effected the minerals sufficiently to cause
serious loss of radiogenic argon.This latter phenomenon is illustrated
by Fig. 2. On this map are shown K-Ar ages on rocks from Eastern N.
America which are all actually of "Grenville" age, i.e., ca. 1000—1100
million years. A considerable body of Rb-Sr and U-Pb age data together with this K-Ar data indicate that these Grenville ages extend

at least from the Labrador coast to Western Texas (TILTON et al., 1960; WASSERBURG et al., 1962; WANLESS et al., 1965; MUEHLBERGER et al., 1964). However the central portion of this region has also been involved in the Appalachian orogeny between 300 and 500 million years ago. The superposition of this younger orogenic belt upon the rocks of the Grenville orogenic belt has resulted in the lowered ages indicated on the figure in the Central Appalachians. In those parts of the Grenville orogenic belt which have not also been involved in the Appalachian orogeny (e.g. in Ontario and Texas) most ages are in the range 900 to 1050 million years and appear only slightly younger than the probable true age of 1000—1100 million years. Extreme cases of this lowering of K-Ar ages as a consequence of a later orogeny have been found in California (LANPHERE, 1962; LANPHERE et al., 1964) in which minerals from Precambrian rocks are found which give Cretaceous ages.

It is also possible to obtain a consistent pattern of K-Ar ages which are nevertheless too low even in the absence of a later orogenic event. This is illustrated by the data obtained on rocks in the vicinity of Baltimore, Maryland, (WETHERILL et al., 1966) which are all at least

Table 1. *K-Ar ages of* ≧ 450 *m.y. rocks from the vicinity of Baltimore, Maryland, U.S.A.*

Sample	mineral	age (m.y.)
Baltimore gneiss (Towson dome)	biotite	340
Baltimore gneiss (Phoenix dome)	biotite	355
Baltimore gneiss (Woodstock dome)	biotite	430
Baltimore gneiss (Woodstock dome)	hornblende	367
Baltimore gneiss (Woodstock dome)	diopside	328
Baltimore gneiss (Woodstock dome)	plagioclase	309
Baltimore gneiss (Hartley Augengneiss)	biotite	290
Gwynns Falls Paragneiss	hornblende	301
Pegmatite in Paragneiss	muscovite	282
Ellicott City granodiorite	biotite	315
Ellicott City granodiorite	hornblende	300
Woodstock quartz monzonite	biotite	295
Kensington diorite gneiss A	biotite	385
Kensington diorite gneiss B	biotite	350
Norbeck diorite gneiss	hornblende	315

450 million years old (Table 1). Some of these rocks are Precambrian in age whereas others are early Paleozoic. The K-Ar ages obtained on minerals from these rocks are almost all about 300 million years, even including measurements on the relatively retentive mineral hornblende. Similar low ages on these samples have also been found by the Rb-Sr method and probably represent a final closure of these minerals at the end of the Appalachian orogeny. This final closure may represent the last of a series of minor metamorphic episodes or may represent the final re-

moval of these rocks from a high temperature environment as a result of epeirogenic uplift near the end of the Paleozoic. Measurement of Rb-Sr and K-Ar ages on Precambrian rocks in Colorado (Wetherill and Bick-ford, 1965; Aldrich et al., 1958) indicates that in this case a time gap of 300 million years intervened between the time of actual emplacement of rocks in an orogenic belt and their final closure. Such time gaps may be a common feature of deepseated plutons and drastically effect the time resolution obtainable in age measurements on such rocks.

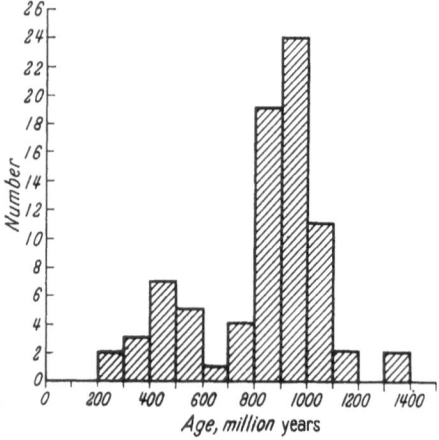

Fig. 3. Histogram showing distribution of K-Ar ages of Norwegian samples (excluding feldspar)

It would be incorrect, however, to give the impression that K-Ar ages on Precambrian rocks invariably leads to confusing and probably meaningless results. The great consistency frequently found in a group of K-Ar ages from a rather large geographical region indicates that under sufficiently tranquil geological conditions radiogenic argon can be retained extremely well, even by biotite, which is more susceptible to such loss than muscovite or hornblende. Two examples of this are given here. The first (Fig. 3) is a collection of all K-Ar data (excluding feldspar) from Norway, based on work of Kulp and Neumann (1960), Faul (1959), Polkanov and Gerling (1960), Gentner and Kley (1957), Goldich (unpublished), and Kley and Schmidlin (unpublished) and discussed by Neumann (1960) which indicates a widespread occurrence of approximately 1000 million years old ages very similar to those found in the Grenville of N. America. Some of the ages in 200—600 million year old range represent rocks which are actually Paleozoic in age, whereas others reflect Precambrian rocks affected by later events (e.g., the Caledonian orogeny). Fig. 4 shows a collection of data from Soviet and Finnish Karelia (Kouvo, 1958; Polkanov and Gerling,

1960; WETHERILL et all., 1962) indicating a similarly widespread occurrence of 1750 million year old ages together with some actual Paleozoic and ~ 2600 million year old ages. Both the ~ 1000 m.y. ages in Norway and the ~ 1750 m.y. old ages in Karelia are well supported by Rb-Sr and U-Pb data. Such agreement of ages cannot be accidental, but it should be remembered that such groups of ages may well include samples of rocks which are considerably older but have been at least mildly metamorphosed during a later regional event and furthermore the

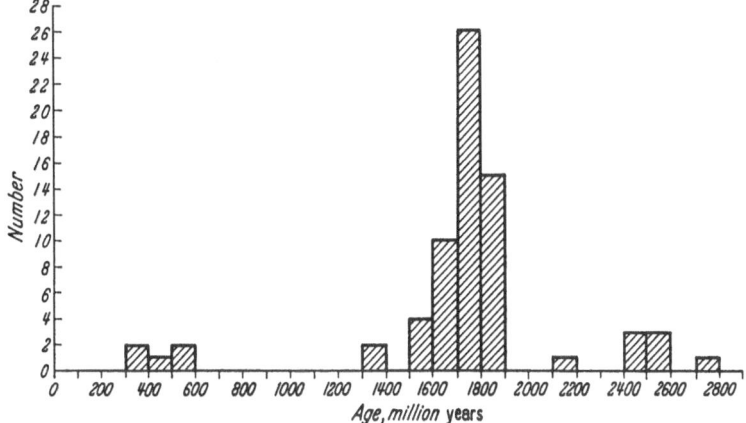

Fig. 4. Histogram showing distribution of K-Ar ages of samples from Soviet and Finnish Karelia (excluding feldspar)

indicated common age of these rocks may well postdate the actual emplacement of even the youngest of the group by up to several hundred million years.

Another approach to K-Ar dating of Precambrian rocks which has received some attention recently is that of dating whole-rock samples of fine grained basic igneous rocks which frequently cannot be dated in any other way. BURWASH et al. (1963) have presented such data for a number of Precambrian intrusives which show a considerable degree of internal consistency and the authors present reasons for believing these ages to be valid. This work, as well as earlier studies of whole rock K-Ar ages of diabase (ERICKSON and KULP, 1961; MCDOUGALL, 1963) indicates that at least in some cases retention of radiogenic argon is nearly complete in the chilled margins of such rocks and also possibly in coarsegrained amphiboles from the central portions of the intrusion. Measurement of whole rock ages of basic rocks containing pyroxene should be regarded with suspicion, however, at least until the circumstances under which this mineral contains excess argon are more clearly understood (HART and DODD, 1962). It may be, as pointed out by BURWASH et al. that such excess argon is found only in pyroxenes of deep-seated origin.

The future of Precambrian K-Ar dating

Several institutions, mostly government geological laboratories, have carried out extensive K-Ar surveys of very large Precambrian areas. The results of such surveys comprise the bulk of the existing Precambrian K-Ar data. Examples of such surveys are the Canadian and Soviet work already mentioned as well as Australian measurements (e.g., RICHARDS et al., 1963). Much valuable information has been obtained by such projects but in view of the foregoing discussion it should also be clear that these results must be used with caution. The greatest part of the Grenville age data for Ontario shown in Fig. 2 is a small part of a very large quantity of Canadian K-Ar age measurements carried out by the Geological Survey of Canada. These results clearly delineate the regions of eastern Canada which were affected by the Grenville orogeny but which were largely unaffected by later events, notwithstanding the fact that there probably are, within this region, rocks which are considerably older. In other cases the interpretation of the K-Ar data is less clear. While there is certainly considerable value in such preliminary studies, a point of diminishing returns is reached after many hundreds of measurements have been made over an area of a million square kilometers. At least by the time this much work has been done it is essential for further progress that the K-Ar work be supplemented by a significant quantity of Rb-Sr studies on whole rocks and separated minerals and if possible by isotopic U-Pb age measurements on zircons. The complexities introduced by the multiply metamorphic history of most Precambrian areas are too great to be unraveled by K-Ar measurements alone; in fact in some cases even all the radiometric and geological techniques available are insufficient to interpret either the regional or local history. Fortunately, however, this is not usually true, but a combined K-Ar, Rb-Sr, U-Pb and geological approach is necessary to resolve ambiguities in interpretation of the data.

For these reasons it may be anticipated that in the future, after these preliminary surveys have been completed, that K-Ar dating of the Precambrian should play a supplementary, rather than a primary, role. There will be occasions when an ambiguity in interpretation of the data can be resolved by K-Ar measurements even when combined with extensive Rb-Sr measurements. This may be illustrated by the data of LANPHERE et al. (1964) in which Cretaceous metamorphism caused the equilibration of strontium between the minerals of a Precambrian rock, and also caused the whole rock samples to be open systems with regard to Rb and Sr. The time of this reequilibration was quite well defined, but the Rb-Sr data alone could not be used to prove the Precambrian age of the original rock. This point could be demonstrated, however, by the 655 million year K-Ar age of a muscovite sample, separated from one of the rocks studied, which had preserved much of its radiogenic argon in spite of the later metamorphism. (Clearer evidence for an age of 1800 m.y. was given by U-Pb data from zircons, however.) It may be expected that in the future the even more retentive mineral hornblende will be used in a similar way in many cases.

Much of this future work will make use of the fact that it can usually be strongly argued that K-Ar ages represent minimum ages of the rock or mineral measured. Radiogenic strontium, and sometimes radiogenic lead, is commonly redistributed between the constituent minerals of a rock during an episode of metamorphism. The Rb-Sr ages of a strontium-rich mineral such as plagioclase feldspar may consequently be apparently too high, owing to the incorporation of this redistributed strontium. This is not usually the case for argon, which under such circumstances is usually lost from the rock, probably to a fluid phase, and ultimately, to the atmosphere. This use of K-Ar data may be illustrated by data obtained by the Geological Survey of Canada (LOWDON et al., 1960, 1961, 1963a, b; WANLESS et al., 1965) for post-Huronian rocks and

Table 2. *K-Ar Ages for post-Huronian rocks in Ontario*

Sample	K-Ar age (m.y.)
Biotite from Nipissing diabase	2095
Whole rock diabase cutting Lorrain formation	1995
Metamorphic biotite from Mississagi quartzite	1625
Whole rock diabase cutting Lorrain formation	1600
Sericitic mica from lower Huronian	1580
Muscovite from post-Cobalt granophyre	1485
Muscovite from Mississagi (?) quarzite	1405
Biotite from lamprophyre cutting Huronian	1395
Muscovite from metaquartzite (Lorrain)	1202
Biotite from olivine diabase	1120

minerals in Ontario (Table 2). Some of these ages probably approximately represent true times of intrusion or metamorphism whereas others are doubtlessly simply too low. However they all are valid younger limits for the age of the Huronian strata. The oldest age (2095 m.y.) brackets the Huronian strata between ~ 2100 m.y. and the 2600 million year old Algoman basement upon which it has been deposited. This result may be compared with the Rb-Sr data of VAN SCHMUS (1965) shown in the form of an isochron diagram (Fig. 5). This whole rock Rb-Sr data was obtained on samples of post-Huronian Nipissing diabase and associated granophyre, and brackets the Huronian between ~ 2150 m.y. and the 2600 million year old basement, in agreement with the conclusion based on K-Ar data.

K-Ar ages would not be minimum ages if minerals contained excess argon, either trapped from a fluid phase during their crystallization or incorporated as a result of subsequent diffusion. The only common rock-forming mineral which has been clearly demonstrated to contain significant quantities of excess radiogenic argon is pyroxene (HART

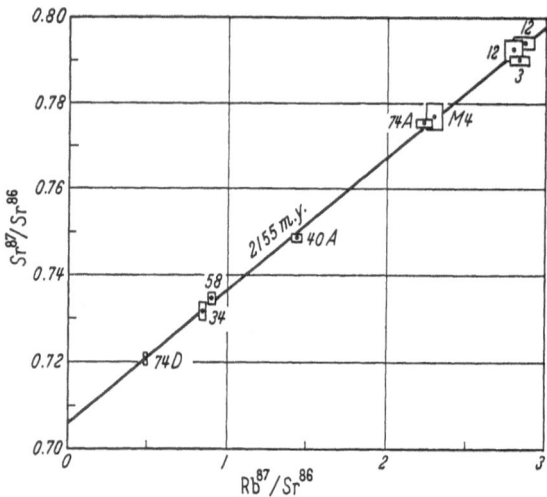

Fig 5. Rb-Sr isochron diagram for whole rock samples of Nipissing diabase (VAN SCHMUS, 1965)

and DODD, 1962). For this reason results of measurements on samples of pyroxene or whole rock samples containing pyroxene must be regarded with suspicion. For example, the data of HART (WETHERILL et al.,

Table 3. K-Ar ages of Baltimore Gabbro

Sample	age (m.y.)
Pyroxene	702
Plagioclase	580
Hornblende	372

1966) shown in Table 3 might be used to draw the important conclusion that the Baltimore gabbro and the Glenarm Series of metasedimentary rocks are of Precambrian age. Such a conclusion, however, would be unwarranted by the demonstrated common occurrence of excess radiogenic argon in pyroxenes. This excess is probably responsible for the very old (up to 6.5 billion years) whole rock ages found by GERLING et al., (1962) on ultrabasic rocks from the Kola peninsula. In this same connection it should be pointed out that the Katarchean rocks of POLKANOV and GERLING (1960) which give K-Ar ages on biotite of ~ 3500 million years also may be an exception to the rule that such ages can be safely interpreted as minimum ages. Recent U-Pb work of ZYKOV et al., (1964) on these same rocks seems to indicate that their age is about 2760 million years. In Fig. 6 are shown the results of this U-Pb work on zircon and monazite from the Murmansk and Voronye river areas when plotted on a concordia diagram. The actual experimental points very accurately define a straight line cutting concordia (WETHERILL, 1956) at 2760 m.y. and also essentially passing through the origin, possibly as a result of U-Pb fractionation during recent weathering. Even including the rather large quoted experimental errors of ZYKOV et al., it is very difficult to interpret these zircon data in terms of the

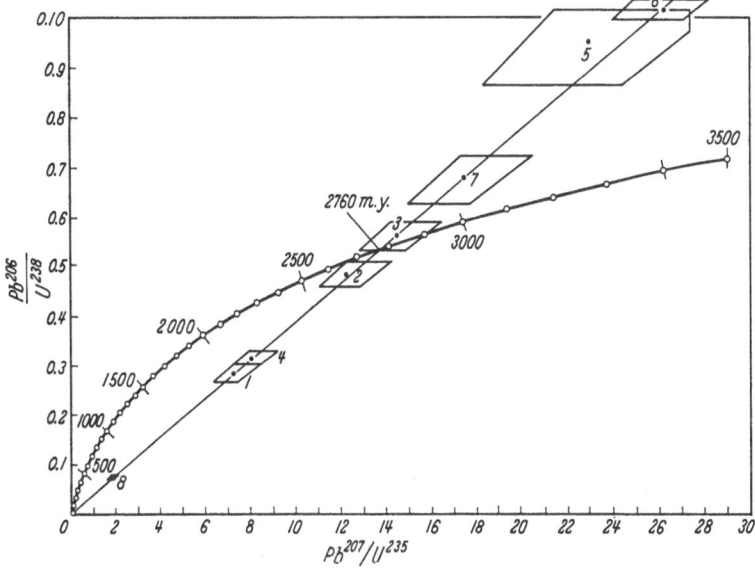

Fig. 6. Concordia diagram showing zircon data in the Murmansk and Voronye river areas of the Kola peninsula, U.S.S.R. (data from ZYKOV et al., 1964)

rocks being 3500 million years old. Such data show that our understanding of the phenomena causing discordant ages is still quite incomplete and only by approaching these problems with a combination of methods can these difficulties be brought clearly to light, and, hopefully, understood.

This work was supported by the National Science Foundation Grant (GP-752).

K-Ar Dating of Plutonic and Volcanic Rocks in Orogenic Belts

By

RICHARD LEE ARMSTRONG

Introduction

Although the radioactivity of potassium was discovered sixty years ago it is only during the last decade that the method of age determination based on the decay of potassium to argon has been widely used. The primary purpose of K-Ar dating has been to determine an age of geological significance.

The large number of K-Ar dates has not always given easily interpreted results but behind apparent contradiction and confusion there must lie a reasonable explanation. In this paper I shall seek to make certain generalizations concerning K-Ar dates and their interpretation in orogenic belts. In spite of the increasing distribution of dates in geologic literature they will never be as important for dating geologic events as fossil age determinations and physical stratigraphy; nevertheless K-Ar dates add a new dimension to our understanding — allowing us to quantify the geologic time scale and to make many previously impossible correlations. The "classical" and radiometric age determinations will interact to lead to a better understanding of geologic history, and ultimately to a better understanding of the underlying forces which have produced that history.

To insure clarity I will use the word "date" when referring to the numerical results obtained on individual samples and "age" when referring to a geologically significant event such as crystallization of a granitic pluton or emplacement of a volcanic rock. Only in certain ideal cases do K-Ar dates exactly equal the ages of events which would be considered geologically significant. These ideal cases and the important exceptions are discussed below.

Conditions for age determination. Accuracy of dates

A known decay scheme, accurate analytical techniques and accurately known decay constants are necessary but not sufficient conditions for a meaningful age determination. In addition the following conditions must be fulfilled:

(1) A K-bearing phase, formed in some "instant" of geologic time, must be available (an "instant" in this case is some span of time less than the error of the time measurement).

(2) At the time of origin this phase must contain no Ar^{40} because no correction is possible for primary Ar^{40} contamination.

(3) Between the time of origin and analysis of the phase in the laboratory the system must remain closed to loss or gain of both K and Ar. This implies a relatively low temperature history and a low diffusion constant for Ar in the substance dated.

Failure to satisfy any one or more of these conditions invalidates the date determined; only where all conditions are fulfilled is a date a measure of age.

·

In addition, experimental error must always be taken into account especially when comparing results from different laboratories. K analyses using flame photometry or isotope dilution are usually accurate to within 2% except for samples with very low K concentrations. As a general rule the accuracy of K-Ar dates should be better than 5% s. d. The problem of accuracy of K-Ar dates has been discussed by BAADSGAARD and others (1957), EVERNDEN and others (1964), ZARTMAN (1964), and DALRYMPLE and HIROOKA (1965).

Materials used for dating plutonic rocks

Given freedom of choice no geochemist would agree to date weathered material. This would be a blatant violation of the requirement of a closed system between time of origin and analysis. Remarkably enough, when weathered samples have been deliberately dated and the results compared with results for fresh material the discrepancies have not been great; this has led to the suggestion that K-Ar dates on detrital material can be used as an indicator of provenance to aid in paleogeographic reconstructions (ZARTMAN, 1964; BAADSGAARD and others, 1964; AFANASYEV and others, 1965; KRYLOV, 1961).

A matter of concern in the collection and preparation of samples is crushing and grinding. Very fine grinding results in distortion and ultimately destruction of the crystal lattice; material for dating should not be ground finer than is necessary for mineral separation. Losses of Ar due to grinding of both micas and feldspars have been reported (GENTNER and KLEY, 1957; GERLING and others, 1960; DOE, 1962) but such losses do not appear to be a general problem. A grain size of 100 mesh is probably a safe practical lower limit; much finer material has been successfully dated, however (BAADSGAARD and others, 1964; EVERNDEN and others, 1964).

Practical experience has shown that not all K bearing substances are equally suitable for dating. Empirical studies comparing naturally coexisting mineral phases and laboratory measurements of Ar diffusion in minerals are in agreement as to the Ar retentivity of the different minerals commonly used for dating. For the dating of plutonic rocks only hornblende and the micas have proven suitable. Plutonic feldspars consistently give Ar results which are from 5 to more than 30% low when compared to micas (FOLINSBEE and others, 1956; WASSERBURG and others, 1956; WETHERILL and others, 1956c; GOLDICH and others, 1957; CARR and KULP, 1957; GENTNER and KLEY, 1957; KRYLOV, 1963; ZARTMAN, 1964). Whole rock samples in which an appreciable fraction of the K is in feldspars are thus unsuitable for dating. In fine-grained, low-grade metamorphic rocks such as phyllites where most of the K is in a mica phase the argon retentivity appears to be quite good and the whole rock sample is suitable as a material for dating (GOLDICH and others, 1957; DODSON, 1963; EVANS and others, 1964; GARETSKY and others, 1964; HARPER, 1964; STEVENS, 1964).

Hornblende has been shown to hold Ar better than the micas when subject to low grade thermal metamorphism (HART, 1961; HART, 1964; THOMAS, 1963; KISTLER and others, 1965). The reasons it is not used more often are its relatively low K content, its scarcity relative to micas in many areas, and the difficulty of obtaining pure hornblende samples. The date measured on a hornblende concentrate with only a few percent mica is effectively a mica date as mica will be the main contributor of K and Ar to the analyses.

Muscovite is perhaps slightly better than biotite in terms of Ar retentivity at moderately elevated temperatures but the comparison of muscovite-biotite pairs often shows no particular tendency of one or the other to give older dates (FAIRBAIRN and others, 1960; WANLESS and LOWDON, 1963; STOCKWELL, 1963). The only important exception is coarse-grained pegmatitic muscovite which, in several cases, has held Ar while nearby finer-grained biotite has lost much or all of its Ar (TILTON and others, 1959; WETHERILL and others, 1960). This is, at least in part, simply a grain size effect on diffusion of Ar at elevated temperatures. Biotite is the mineral most widely used for K-Ar dating because of its wide range of occurrence, ease of concentration, high K content, and its high Ar retentivity. On all micas both K-Ar and Rb-Sr dates may be determined. This provides a useful check although it is known that even concordant results are not necessarily a significant measure of age.

One of the basic conditions for K-Ar dating is that the mineral phase dated contained no primary Ar^{40} at the time of its origin. This is never strictly true. In the natural environment, particularly at great depths within the crust, excess Ar^{40} is present in whatever fluid phases exist. During remobilization of an ancient metamorphic terrane quite high Ar pressures might develop. No mineral phase ever crystallises absolutely free of contamination from its environment; this contamination may occur on an atomic scale with foreign atoms being accidentally trapped in the crystal lattice, or as bulk contamination in the form of solid and fluid inclusions. It is only logical to accept that a finite Ar^{40} background must exist for every mineral. The practical question is to what extent this background affects mineral dates.

Analyses of K deficient minerals have revealed Ar^{40} excesses in beryl, cordierite, tourmaline, apatite, fluorite, sodalite, albite, quartz, and pyroxene (DAMON and KULP, 1958; LIPPOLT and GENTNER, 1963; YORK and others, 1965; DAMON and others, 1964; RAMA and others, 1965;

HART and DODD, 1962; GERLING and others, 1962; HART, 1963; MC-
DOUGALL and GREEN, 1964). The excess is particularly notable in pyroxenes
of very low K content. Pyroxene from hypabyssal plutons has been
reported to yield satisfactory dates (McDOUGALL, 1963c); more complete
degassing might be expected in low pressure environments so this observation is not in conflict with the occurrence of excess Ar in pyroxene
from deep-seated plutonic environments. Most studies of whole-rock
dates on basic dikes have not encountered excess Ar^{40} but in at least one
case excess did occur (ERICKSON and KULP, 1961; BURWASH and others,
1963; FAHRIG and WANLESS, 1963; MILLER and MUSSETT, 1963; STOCK-
WELL, 1963; BAADSGAARD and others, 1964).

Most serious is the possibility of excess Ar^{40} in the minerals most
commonly used for dating; hornblende and mica. Examples of excess Ar in hornblende seem to exist (LEECH and others, 1963; HUNT,
1962) and it is an open question as to whether some of the cases of better
Ar retention reported for hornblende actually represent compensating
effects — excess Ar^{40} cancelling out the effect of Ar loss.

Most available information supports the conclusion that Ar^{40} excesses do not occur in micas to a sufficient extent to alter the dates
determined (DAMON and KULP, 1957 b). Remarkable levels of contamination are required to affect mica dates but experimental results exist
which show that Ar may be incorporated into mica in large quantities
so completely that it is only extracted at high temperature and thus
inseparable from radiogenic Ar (KARPINSKAYA, 1961 and 1964; PEPIN
and others, 1964). Perhaps at very high pressure Ar no longer behaves
as an inert gas.

We do have geologic evidence that there may be excess Ar in mica.
The most convincing case comes from the regional studies of the Geological Survey of Canada (STOCKWELL, 1963). A biotite with 8% K gave
a date of 3.7 billion years for a locality along the Grenville front where a
date between 1 and 2.5 billion years would be expected. Either this
represents excess Ar^{40} in mica or an entirely new orogenic belt preserved
intact in a single locality and unsubstantiated by any other dating method
must be postulated. The first explanation appears less contrived, although
more upsetting. Other results for profiles through the Grenville front
give results suggestive of excess Ar in biotite but the conclusion is not as
compelling.

The conclusion that excess Ar can be found in all dating materials
is sobering. For the important dating minerals hornblende and mica the
chances of obtaining a misleadingly large date are slight — certainly not
enough to discredit K-Ar dating but great enough to require caution in
the interpretation of individual results.

Materials used for dating volcanic rocks

Volcanic rocks are eminently suitable for K-Ar dating. Hornblende and biotite are common in many types of volcanics and even where they do not occur suitable material for dating may be present. Most important are the high-temperature feldspars, the best and most completely studied of these is sanidine. Experimentally and empirically the Ar retentivity of sanidine is close to that of biotite (FOLINSBEE and others, 1960; BAADSGAARD and others, 1961; EVERNDEN and others, 1960, CURTIS and others, 1961; FECHTIG and others, 1961; EVERNDEN and others, 1964). Anorthoclase and plagioclase have also been successfully employed for dating volcanics (McDOUGALL, 1963c; EVERNDEN and others, 1964). Pyroxene has also been used with reasonable success (McDOUGALL, 1963c). Excess Ar has not yet been reported for phenocrysts in volcanic rocks, but it has been found in xenoliths enclosed within volcanics (DALRYMPLE, 1964; EVERNDEN and others, 1964); xenoliths should be avoided, if possible.

Volcanic glass has been studied and found promising for dating if completely unaltered — a difficult matter to decide (SCHAEFFER and others, 1961; EVERNDEN and others, 1964). Glass appears to give good minimum dates and provides a useful check on dates for coexisting minerals. Used alone to date volcanics it is not particularly reliable; many results from suitable appearing material have to be discarded as meaningless. Similar mixed success has been encountered in the dating of whole-rock samples of volcanics (ERICKSON and KULP, 1961 and 1964; Mc-DOUGALL, 1964; EVERNDEN and others, 1964; LIPPOLT, 1961). Fresh fine-grained samples ranging from rhyolite to basalt have produced acceptable results; low values are sometimes found, however, and there appears to be no objective means for predicting the behavior of a given sample — except for the case of those samples which are obviously altered, and not worthy of further consideration. The best policy is to obtain enough dates on differing stratigraphic units and types of material so that discrepancies can be recognized and eliminated.

Calibrating the stratigraphic record

It is *only* in the case of volcanics and hypabyssal plutons that the basic assumptions of K-Ar dating are rigorously satisfied. Dates on metamorphic rocks and authigenic minerals in sediments supply minimum estimates of age and are unsuitable for precise definition of the geologic time scale.

Volcanic and hypabyssal plutonic rocks do cool and crystallize in an instant of geologic time. Minerals in volcanic rocks, if not subsequently altered, do retain argon; problems with excess Ar^{40} in volcanic minerals have not been encountered. Volcanics are abundant and accessible, and are capable of precise stratigraphic correlation and age determination; single units may supply several mineral phases as well as unaltered glass for checks on the internal consistency of the dates measured. The most exhaustive, elegant, and successful of the attempts to calibrate the time scale absolutely are those of EVERNDEN and others (1964) and FOLINSBEE and others (1961).

The time scale now being used is the one reviewed by KULP (1961). Control for the entire scale rests on accurately determined dates on volcanic material intercalated into the stratigraphic record and a few favorable cases where dated hypabyssal plutons are closely bracketed stratigraphically. The time scale represents an important achievement of K-Ar dating. No other method is so easily applied and suitable for dating volcanic rocks, the effective range beginning with ages of less than 1 million years and covering the rest of the time scale. The process can, and has been, reversed — dating of volcanic and hypabyssal plutonic rocks of unknown stratigraphic age; such studies represent a major contribution of K-Ar dating to geologic knowledge.

Interpretation of dates in an orogen

To begin the discussion of orogenic belts it is best to confine the discussion to dates determined on medium to fine grained micas — the information most commonly available. Exceptions encountered by dating hornblende and pegmatitic mica can be better discussed after a general model is proposed, To review the masses of available data and derive the generalizations concerning their interpretation would be logical, but much more efficient is the presentation of a simplified, although not necessarily simple, model from which the generalizations may be derived. Then, by means of an example, it can be shown how the real results can be interpreted in terms of the model.

The model orogen

At this point I wish to introduce the "model orogen", a simplified orogenic belt drawing its manifold features from examples in the real world. The model reflects my knowledge of the Cordillera and Appalachians in North American and the Alps and Caledonides in Europe. Fig. 1 is an attempt at a representative cross section of the model orogen in its present structural state and a graphical representation of dates and ages within the orogen. A major but not invalidating feature of the model is that it has gone through a single long orogenic cycle. The geologic evidence might be interpreted, however, as indicating

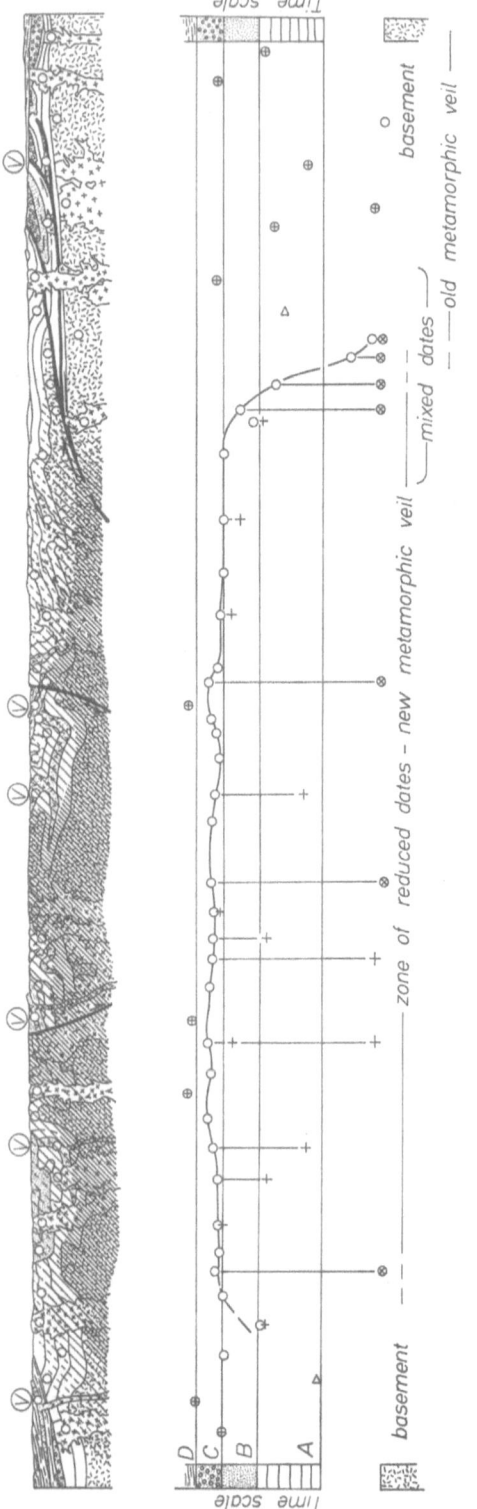

o Sample location on profile
o Date determined on fine to medium
   grained mica
+ Time of intrusion or extrusion
△ Date on detrital or authigenic mica
Ⓥ Volcanic sample
⊗ Date on older metamorphic veil
   before remetamorphism

D Postorogenic period
C Late synorogenic-early postorogenic period
B Early synorogenic period
A Preorogenic period

Chlorite isograd
Biotite isograd
Sillimanite isograd

basement — · — · — zone of reduced dates – new metamorphic veil ⎯⎯⎯⎯⎯ mixed dates ⎯⎯⎯⎯⎯ old metamorphic veil ⎯⎯⎯⎯⎯ basement

Fig. 1. Profile of Geochronometric model orogen illustrating the metamorphic veil

several discrete orogenies within this orogenic cycle — it is a matter of philosophy and semantics, and not a dispute over what actually happened. The history of the model orogen begins with an ancient crystalline basement with cross-cutting granitic plutons. Upon this basement a thick sequence of sediments is laid down during preorogenic period A. Volcanic rocks occur locally within this sequence of sediments. The deformation of the orogen begins in more central zones; folding, pluton emplacement and metamorphism occur over a long period of time, period B, the early synorogenic or "flysch" period. Sediment accumulates in basins within the orogen concurrently with deformation, local relations within such basins might be interpreted as representing two or more distinct orogenies — one for each angular unconformity — but the overall picture is not so simple as deformation was occurring more or less-continuously with time within the orogen as a whole — only locally deformation appears episodic. During the B period the orogenic belt appears as a sediment source for areas outside the region undergoing deformation. Plutons emplaced at the end of period B within the interior of the belt appear structurally post orogenic. They crosscut metamorphic fabrics and large scale structures in metamorphic rocks. By the end of the period the interior of the orogen is largely consolidated.

Period C is the late synorogenic to early post orogenic or "molasse" period. Perhaps as early as the latter part of period B deformation begins along the margin of the orogenic belt; in the transition zone from craton into geosyncline a fold and thrust belt begins to form. The main evolution of this fold and thrust belt, however, is during period C. Great overthrusts with tens of miles of displacement transport geosynclinal rocks out over the craton. Large quantities of coarse debris, derived from the evolving fold and thrust belt flood the marginal troughs, filling them and driving out the sea irrevocably. This sedimentation is concurrent with the thrusting. Clasts derived from the growing thrusts can be recognized in the coarse debris overlying and overridden by the thrust plates. Locally within the orogen volcanism and plutonism occur during period C.

Period D is the postorogenic or "block faulting" period. Compressive deformation within the orogenic belt has ceased. Block faulting and warping produce basins in which postorogenic sediments and volcanics accumulate. Associated with the volcanics are postorogenic plutons. Except for rare unpredictable alkalic plutons the geologic history of the model orogen is complete. The model may seem contrived or in dispute with extant interpretations of orogenic histories but I feel it closely resembles actual examples. The model serves its purpose by providing examples of all the most commonly encountered situations where dating might be applied.

The "metamorphic veil"

The "metamorphic veil" is a conceptual surface in space-time that synthesizes the dating results on metamorphic rocks. The "veil" is indicated by the bolder line on the graphical representation of dates on Fig. 1. I first ask the questions: what do we mean by age of metamorphism and exactly what are we dating when we date a metamorphic rock? Answers to these questions lead us logically to the "veil" concept. Detailed and exhaustive syntheses of structural and petrographic data, particularly those

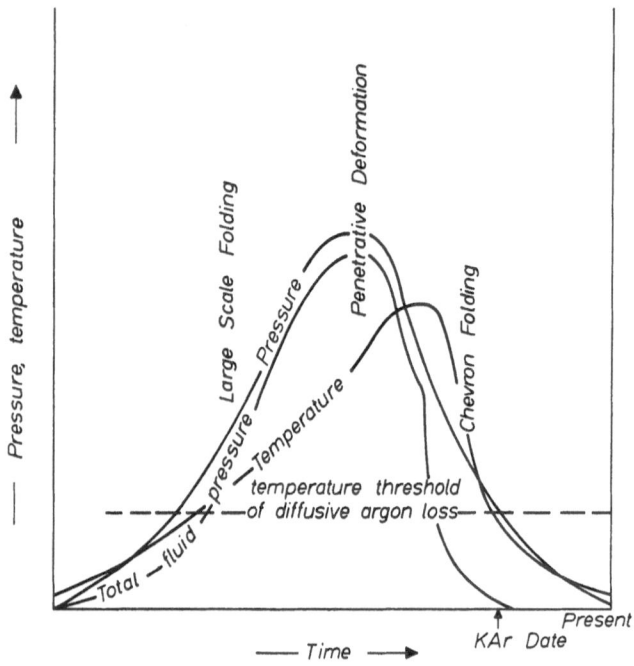

Fig. 2. Temporal relationships of temperature, pressure, and deformation in the metamorphic interior zone of an orogenic belt. The point in time given by a K-Ar date is indicated. Diagram modified from SCHUILING, 1963; DEN TEX, 1963; JOHNSON, 1963

of European geologists, lead to the idea that have shown a certain normal progression of events occurs in a metamorphic terrane. Fig. 2 is a graphical representation of this progression of structural and metamorphic events for a given area. There is no exact correspondence between structural events and the growth of metamorphic minerals. During early stages of increasing metamorphic grade, large — scale recumbent folds develop. This is accompanied and followed by development of penetrative structures — minor folds, lineations and so forth. During the progression of metamorphism volatiles are continually produced by dehydration reactions and lost from the overall system. The volatiles catalyse flowage in the rocks; their loss results in increased viscosity; when the maximum metamorphic grade is attained, most of the deformation shown by strain (folds etc.) in the rocks has ceased. Quite commonly the index minerals for the highest grade reached (almandine amphibolite facies is the normal maximum encountered in younger orogenic belts) have grown postkinematically. Deformation following the temperature maximum is usually a more nearly "brittle" chevron folding, if any. As the temperature falls from its maximum the rock suddenly becomes rigid and inactive — any volatiles remaining are exhausted by the first infinitesimal reactions of retrogression. Unless volatiles are reintroduced along shear zones retrogression ceases; the dry systems are static. Because retrogression reactions are commonly hydrations involving volume increase they are self — inhibiting — as volatiles begin to move in, the volume increases of the retrogressive reactions immediately seal the passageways of access and further reaction stops. This is exactly the converse of progressive metamorphism — progressive reactions involve volume decrease and release of volatiles — thus during progression there is juice to spare for catalyzing reactions and recrystallization flow and at the same time passageways for escape of volatiles are created. It is thus not difficult to understand why progression occurs so much more rapidly and completely than retrogression during regional metamorphism. Cooling after the thermal maximum proceeds more or less irregularly until the rocks reach the surface of the earth where they are available for our observation and collecting for dating.

The question of what is meant by "age of metamorphism" is not easily answered. Much geologic thinking is vague on this point — usually the whole process is considered to occupy only an instant of geologic time, an "orogeny". If this were true there would be no problem but studies of younger orogenic belts have shown that it is definitely not so — the progression of deformation and metamorphism and cooling require millions of years at a minimum and may involve tens of millions of years!, certainly a significant span of time. We are forced to conclude that *"metamorphism" has no exact age — it spans a period of time*, a period which may be difficult to interpret and define but nonetheless a span of time for which no single age applies! What, then, do we date ?

We know from experimental and geochronometric studies that micas will loose argon within geologically short periods of time at temperatures as low as 200° C (HURLEY and others, 1960 b). Thus we are determining an approximate time when the rock system cooled through some critical isotherm; daughter retention begins over some unspecified time interval in the cooling history of the rocks.

To speak of an isotherm is undoubtedly an oversimplification; factors such as pressure, shear, presence of volatiles, and chemical composition of the system certainly affect the exact temperature range over which the minerals freeze in their radiogenic Ar, but temperature is certainly a primary control and an adequate conception for our model. We must thus inquire when, in relation to the span of time of metamorphism, this critical isotherm was crossed — years or millions of years thereafter. The higher grade zones certainly must start out at least 200° C above the critical isotherm. Cooling a warm body the thickness of the crust by conduction through its upper surface only is not a rapid process. At a minimum, millions of years are probably involved. If the rock system remains at the depths where it was metamorphosed it may never cool below the critical isotherm; the rocks we observe and date however have been uncovered by erosion. The time of passage through the critical isotherm is more closely related to the erosion history of the region than to the time when the thermal maximum was reached! The close correspondence of frequency maxima of dates and rapid erosion in the Appalachians has been pointed out by HADLEY (1964).

Many mineral dates in the Alps are closer to today than they are to the time of deformation in the metamorphic zones (JÄGER, 1962). In the western U.S. many dates on metamorphic rocks are closer to today than to the time of their metamorphism and deformation (ARMSTRONG and HANSEN, 1966). The dates are in some cases nearly 100 million years younger than the end of the time span of metamorphism and deformation as indicated by the ages of postorogenic plutons.

The only logical conclusion is that no method of radiometric dating can give the exact age of development of structures and fabrics in metamorphic rocks. The dates most commonly measured are always less than the time when the metamorphic culmination was reached, probably by millions to hundreds of millions of years. To speak of dates as a measure of the "age of metamorphism" is manifestly incorrect. These dates supply no more information than that classically supplied by the relationship of an unconformity — an upper limit to the time of metamorphism. In many areas, classic geologic criteria for dating of metamorphism and deformation — based on provenance of derived sediments, unconformities, cross-cutting relationships such as plutons and dikes, and more indirect evidence have provided more accurate age determinations than the radiometric methods.

We thus return to the "metamorphic veil", a surface in space-time — a contourable surface with years as a unit of measure. When enough information becomes available, a contour map for the orogen can be made which represents the time when the various portions of the orogen now observed at the surface passed through the "critical isotherm". The contour lines may be termed "chrontours" if a distinctive name is considered useful. All samples, including those from plutons and volcanics and even "postorogenic" plutons, that are older than the local veil, will give dates defining the veil. Evidence of a previous history will be destroyed as far as the finer-grained mica dates are concerned. Only dates on samples outside the metamorphic veil will measure ages; ages for volcanics and plutons much younger than the veil will be correct even in the core of the orogen. Ages for postorogenic, synorogenic, and preorogenic plutons will be determined only in areas away from the zone of metamorphism or where the veil is distinctly older than the age being determined.

It might be suggested that the mean or mode of the date histogram gives a significant number but I see no reason why this should be so. The closest dates to the metamorphic culmination should be those from low grade metamorphic zones. Perhaps there is a sort of plateau on the veil as suggested on the diagram — a plateau which on the edges of the orogenic belt, most closely approximates the time of metamorphic culmination. There is, as yet, no proof that such a plateau exists; it is equally probable that the transition from dates representing metamorphic minerals to those which represent mixed authigenic detrital material is completely gradational and that no inflection in the veil can be recognized. We are left with the reality that *we do not yet know any way of unequivocally and precisely dating any instant in time related to the time span of metamorphism.*

Many of the fine distinctions of interest in the dating of younger rocks are unattainable in the study of Precambrian rocks. To be resolved, date groups must be separated by at least 50 m.y. at 1000 m.y., and by 100 m.y. at 2000 m.y. The study of the geochronology of Minnesota by GOLDICH and others (1961), where limited resolution of date groups within major orogenic cycles was achieved, is one exception; most dating studies have been unable to do more than map out the distribution of orogenic belts of distinctly different age.

K-Ar dating of Precambrian rocks can outline the "metamorphic veil" of ancient orogens. The dates obtained do not give the age of the ancient orogeny as is sometimes stated — they are merely minimum values for the true age of the deformation and metamorphic culmination evidenced by the rocks. The "magic numbers" provide identity tags for the rocks, allowing geologically useful subdivisions to be made. The orogenies dated provide a means for a broad-scale subdivision of Precambrian time, but the dates are only approximations of the absolute ages involved.

Ages may be determined for exceptional areas where Precambrian volcanics, hypabyssal plutons, and dike swarms are preserved in an unaltered state. The dating of basic dike swarms to define significant instants in Precambrian time has only begun and may eventually prove to be an important factor in refinement of the Precambrian time scale (FAHRIG and WANLESS, 1963).

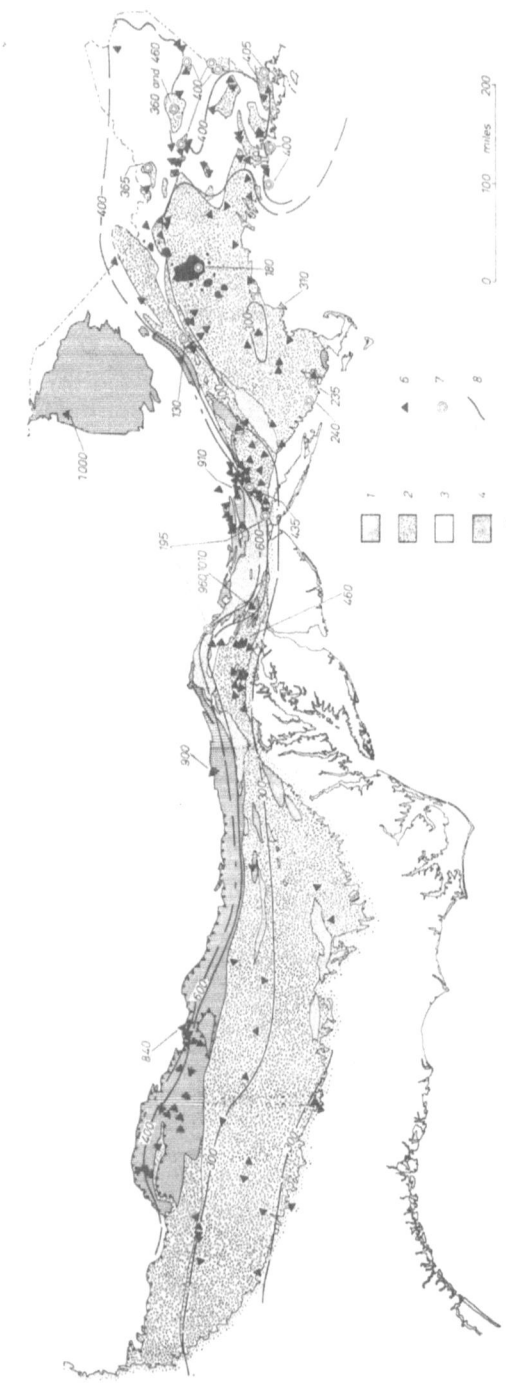

Fig. 3. Map of the Appalachian region summarizing the results of K-Ar dating studies. Explanation of symbols: 1 Precambrian crystalline rocks; 2 Paleozoic plutons and metamorphosed Paleozoic rocks; 3 unmetamorphosed Paleozoic; 4 Triassic basins; 5 Cretaceous and Tertiary coastal plain overlap; 6 K-Ar date locality; 7 K-Ar date locality where date is an age determination; 8 chrontours; 9 plutons of the White Mountain Magma series. Map modified from U.S.G.S. — A.A.P.G. Tectonic map of the United States (1961)

Penetrating the veil

It is obvious that K-Ar dating is not the way to learn all there is to know about the history of an orogen. K-Ar and Rb-Sr dates on finer grained mica both record the "metamorphic veil", but there are ways to peer farther back into a rock's history. Whole rock Rb-Sr isotopic studies and zircon dates are undoubtedly the most direct attack on this problem but they are not the topic of this paper.

There are limited possibilities of penetrating the "veil" using K-Ar dating alone. Very coarse muscovite from pegmatites has, in some cases, retained Ar much better than micas in country rocks (TILTON and others, 1959; WETHERILL and others, 1960). A much closer estimate of times of crystallization and of the age of the metamorphic culmination is thus possible — but it is still only a *minimum* estimate. Similarly the greater Ar retentivity of hornblende can be utilized to obtain dates closer to the ages of geologically significant events; complications may be encountered, however, if incompletely degassed relict hornblende or hornblende with excess Ar are present.

It is also possible in exceptional cases to find isolated masses of rock which have been less affected by regional metamorphism — most significant in this respect are large, homogeneous, dry, unsheared masses of basic igneous rocks, which may yield minerals which give much older dates than the surrounding country rock (examples may be found in POOLE and others, 1963; FAUL and others, 1963; LAPHAM and BASSETT, 1964). The survival of primary igneous textures in such masses of basic rock is well known and provides a clue in the search for these date "asylums".

K-Ar dates in the Appalach ian orogen

The interpretation of dates from an orogenic belt is a matter of deciding which dates to assign to the "metamorphic veil" and which may be taken as meaningful estimates of geologic ages. The veil is not always easily recognizable; GABRIELSE and REESOR (1964) discuss the confusing assortment of K-Ar dates obtained in two large regions in British Columbia, Canada. Even by making maximum use of known structural and stratigraphic relationships the results could not be unambiguously interpreted in every case; no simple metamorphic veil will suffice to explain all the remaining discrepancies. The Appalachian orogen in the United States presents a much simpler collection of K-Ar dates, perhaps because it has been more deeply eroded than the Cordilleran orogen in British Columbia. Fig. 3, based on the data sources in Table 1 summarizes the K-Ar results for the Appalachian orogen; the important orogenic movements in the region are listed in Table 2.

Most of the available dates have been interpreted as belonging to the Appalachian metamorphic veil; the flatter portions of the veil lie between 400 and 250 m.y. In many areas the veil date is not in accordance with the age of deformation and end of metamorphism as interpreted from geologic criteria.

In the southern Appalachians no distinct plateaus on the veil exist and few specific conclusions, except an upper limit for the age of metamorphism in the central Piedmont, can be reached. A transition zone

Table 1. *Sources of data for figure 3, the map of the Appalachian region summarizing the results of K-Ar dating studies*

DAMON and KULP (1957)	KULP and ECKELMANN (1961)
DAVIS and others (1962)	LAPHAM and BASSETT (1964)
DOE (1962)	LONG (1962)
ERICKSON and KULP (1961)	LONG and KULP (1962)
FAIRBAIRN and others (1960)	LONG and others (1959)
FANALE and KULP (1962)	TILTON and others (1958)
FAUL and others (1963)	TILTON and others (1959)
HART (1961)	TILTON and others (1960)
HURLEY and others (1958)	WASSERBURG and others (1956)
HURLEY and others (1960)	WASSERBURG and others (1957)

Table 2. *Summary of orogenic movements in the Appalachian region after* RODGERS (1966)

Palisades — Late Triassic
Alleghanian — Pennsylvanian and/or Permian
Acadian — Middle Devonian to Mississippian
Salinic — Late Silurian (local)
Taconian — Middle and Late Ordovician (locally older)
Unnamed — Early Ordovician or older (local)
Latest Precambrian
Grenville — Late Precambrian

from Paleozoic to Precambrian dates is evident along the western margin of the veil; these dates have no particular geologic significance. In New England uneven plateaus on the veil are present at approximately 360 and 250 m.y.; these dates would correspond with the Acadian and Alleghanian orogenic movements but the exact relationship is unclear. Areally, each plateau includes regions that were deformed significantly earlier than the date given by the plateau as well as regions which were deformed immediately previous to it. The lower the grade of regional metamorphism associated with a given area, the closer its plateau date corresponds with the age of the deformation. This is what would be expected; along the edges of metamorphic belts dates are better estimates of ages of deformation than elsewhere.

In northern Maine the regional metamorphism dies out and the veil plunges northeastward below 400 m.y. and becomes diffuse and erratic. In this area dates on granitic plutons appear to approximate the ages of the plutons, which cluster around 360—365 m.y. and 400—405 m.y. (FAUL and others, 1963; HURLEY and others, 1958). These plutons would be associated with the Acadian orogeny in Middle Devonian time and the Salinic or an early Acadian orogenic activity close to the beginning of the Devonian. The Cortland complex (435 m.y.), on the edge of the veil near New York City, is the only other example of a Paleozoic pluton older than the veil for which the K-Ar date appears to be a reasonably reliable age determination (LONG and KULP, 1962). In the Appalachian region of Canada where regional metamorphism is less intense and the metamorphic veil less important many of the dates determined have been useful age determinations; the results for this region have been reviewed by POOLE and others (1964).

Slightly to distinctly younger than the veil are the late Paleozoic granites in Rhode Island (235—240 m.y.), the alkalic plutons of the White Mountain magma series (130—180 m.y.), and the diabase intrusives in the Triassic basins of New York, New Jersey and Pennsylvania (195 m.y.). These dates are probably all reasonable estimates of age.

Only a few results penetrate the veil. Most significant is the work of HART (1961). His dates on hornblende from Precambrian basement in New York, North Carolina, Virginia and Pennsylvania (840—960 m.y.) have been older than dates on associated fine grained mica but even so they may have been affected by Paleozoic metamorphism and may not be real age determinations. A date of 310 m.y. reported by HART for a granitic pluton near Rockport, Massachusetts is older than the metamorphic veil in the same area but its exact significance is unknown.

Two examples where micas associated with basic rocks have given dates distinctly older than their surroundings are the gabbro near Mount Katahdin, Maine (460 m.y.) dated by FAUL and others (1963) and Southeastern Pennsylvania plagioclase hornblendite (460 m.y.) dated by LAPHAM and BASSETT (1964). These dates may be ages but there is no proof in either case; Nevertheless they clearly restrict the ages of the associated plutons much better than dates associated with the metamorphic veil. The oldest K-Ar dates available for Precambrian basement are 1010 m.y. for biotite from southeastern Pennsylvania (TILTON and others, 1960) and 1000 m.y. for coarse muscovite from the western Adirondacks (DOE, 1962). These are the best K-Ar determinations for defining the upper limit of the age of the Grenville orogeny in the Appalachian region.

Much more is known of the geochronology of the Appalachian orogen as a consequence of Rb-Sr and U-Pb dating than is presented in this review of the results obtained solely using K-Ar dates. The K-Ar dates enable us to define a metamorphic veil for the region and to pick out certain areas above or outside the veil where such dates represent real age determinations. In a few cases the veil has been at least partially penetrated, but the important events behind the veil are probably discernable only utilizing the full range of geochronometric methods.

K-Ar Dating of Sediments

By

P. M. HURLEY

Introduction

It is particularly appropriate to review the subject of K-Ar dating of sediments in a volume that honours Professor GENTNER. Much of the earliest work on the testing of the build-up and possible escape of Ar^{40} in K-bearing minerals of known age was carried out in his laboratory. Evaporite minerals from sedimentary sequences were chosen because of high content of potassium and well-established geologic age. The studies involved the branching ratio in the K^{40} decay scheme, the mechanism of argon loss and diffusivity constants, as well as the suitability of minerals for dating. These subjects subsequently became areas of extensive research by other laboratories.

Historically the geologic time-scale has been related to periods of sedimentation. It is therefore natural that considerable attention should be paid to the problem of dating times of sedimentation, as this would most directly tie into the most abundant and well-classified information on geologic history. However, the dating of the time of sedimentation is not the dating of sediments, because for the most part, sediments are materials that had a time of origin prior to the period of sedimentation. This leaves only the possibility of dating truly authigenic materials which became closed systems at the time of sedimentation. Such materials are few, and may include inorganic and biogenic carbonates, evaporites, and marine minerals such as glauconite, phillipsite, illite and a few others.

In considering the various possibilities it is well known that most carbonates of biogenic origin gradually recrystallize or exchange with their environment, thereby failing in the requirement of a closed system. The evaporites similarly are easily recrystallized in the presence of even trace amounts of water, and the minerals in the system are constantly tending toward new equilibria as a bed of evaporites is buried to greater depths. Authigenic iron minerals that are characteristic of iron formations of marine origin, are too infrequent in occurence in most sedimentary sequences so that they are eliminated as possibilities for general use. Also most of them are too low in a suitable parent nuclide, so that the measurement of isotopic ages becomes difficult. The outstanding exception to this is, of course, glauconite.

The earliest investigations, however, were on sylvite. The first attempt to demonstrate the age relationship of the Ar^{40}/K^{40} ratio in sedimentary minerals was performed by ALDRICH and NIER (1948) on sylvite and langbeinite. Although they did not actually calculate mineral ages in this work, they foresaw that the K^{40}-Ar^{40} decay might become useful in the measurement of geologic time. INGHRAM, BROWN, PATTERSON, and HESS (1950) investigated a sylvite sample from Permian evaporite deposits at Stassfurt. The age that can be calculated from their date is too young for the Permian, suggesting that the mineral was recrystallized, with attendant loss of Ar^{40}. During the early 1950's GENTNER and coworkers experimented with the dating of sylvite from evaporite deposits of the Lower Oligocene of the upper Rhine, and found a direct relationship between argon content and crystal size, the variation being shown to be due to a bulk diffusion process. (SMITS and GENTNER, 1950; GENTNER et al., 1953a, 1953b, 1954b). They concluded that the salt layers were formerly at a higher temperature than at present and calculated an argon diffusion coefficient to apply in determining the absolute age of the deposits. Shortly after these investigations MOLJK, DREVER, and CURRAN (1955) obtained a date for sylvite from the Stassfurt deposits that was younger than expected, and they pointed out that for accurate potassium-argon dating of minerals appreciable amounts of argon must not have escaped from the mineral, and the branching ratio and half-life must be known. The obvious problems in the use of evaporite salts for dating sediments soon led investigators to try more stable minerals. It was natural that glauconite would be highest on the list of possibilities.

The study of glauconite

Glauconite contains abundant potassium which makes it particularly suited to the K-Ar method. It is distributed widely enough so that it is obtainable in sequence of geologic ages since the late Precambrian. In its favor is the fact that it is a low temperature mineral, so that the very presence of glauconite in a sediment suggests that its history of heating has not been great. This means that the diffusion loss of argon was at least low at the start, as opposed to the necessarily elevated temperatures and therefore argon loss in the case of potassium minerals forming in igneous rocks. However, the advantage of the relatively low temperature history is offset by the likelihood of partial recrystallization to a more stable form under increasing temperature and pressure, as the sediment is buried. Finally there is the possibility of initial argon when the glauconite develops in a clayey environment as the result of the occlusion of micas or other detrital impurities.

The question of inherited argon by the admixture of other detrital minerals that are rich in potassium can be investigated fairly simply by successive purifications of the glauconite sample. Any observed decrease in the K-Ar age may be used to estimate the magnitude of any error due to this effect, and samples may be selected on this basis. The possibility of recrystallization or readjustment of the glauconite to new environments, and concomitant gain of parent or loss of daughter, is much more difficult to evaluate. A number of investigators have attempted to measure argon diffusion loss from glauconite at elevated temperatures in the laboratory. These investigations are discussed in the chapter on argon diffusion. There are evidences that argon loss from glauconite may not be due entirely or at all, to normal processes of diffusion. It may rather be due to reconstruction of the structure which apparently involves a slight uptake of potassium during a slow process of diagenesis. In considering these uncertainties it is best to look first at the actual measurements on natural samples of glauconite that have been collected from various stratigraphic columns over the earth, from different levels or depth zones, and of different ages.

Historically, the earliest attempts to test the K-Ar age of glauconites of known geologic age were made by FOLINSBEE, LIPSON and REYNOLDS who reported separately and together in 1956. (FOLINSBEE et al., 1956; LIPSON, 1956; REYNOLDS, 1956b.) They measured twelve sedimentary mineral samples ranging in apparent age from 16 to 285 my. including ten glauconites, a sylvite, and a feldspar from a volcanic tuff, and they correlated all sample results with the Holmes B time-scale. They suggested that the age discrepancies might be due to Ar[40] inheritance and to loss through weathering. In the same year WASSERBURG, HAYDEN, and JENSEN reported on three glauconites and an authigenic feldspar and also suggested that authigenic minerals may inherit radiogenic argon from the pre-existing minerals which they replace. (WASSERBURG and HAYDEN, 1956; WASSERBURG et al., 1956.) The uncertainty in time after sedimentation at which authigenic minerals may have formed, was mentioned as a problem.

In 1957 ORCHINNIKOV, SHUR, and PANOVA (1957) reported on argon age determinations on sedimentary rocks from the Urals. AMIRKHANOV et al., (1957), (1958a) presented age determinations of many glauconites from various parts of Daghestan. AMIRKHANOV et al. (1957) found from heating tests that argon is lost from glauconite more readily than from mica or feldspar, and noted that paleotemperatures, metamorphism, local heating, and weathering may be important considerations in using glauconite for dating. SARDAROV (1957a), (1958) concluded that glauconite was generally suitable for age determination by the K-Ar method.

Starting in 1958, a number of laboratories issued reports of continuing studies on glauconite and other sedimentary materials. AMIRKHANOV, BRANDT, BARTNITSKIY, GURVICH, and GASANOV (1958) made thermal studies of glauconites to determine the degree of retention of radiogenic argon. KAZAKOV and POLEVAYA (1958) dated glauconites from Sinian to Upper Eocene, and compared these results with other methods of age determination. POLKANOV and GERLING (1958) reported a glauconite date from the Sinian of China. CURTIS and REYNOLDS (1958) and LIPSON (1958) discussed potassium-argon dating of sedimentary rocks, and pointed out possible sources of errors due to contamination of older material imbedded in glauconite, argon inheritance, argon loss by diffussion, the presence of atmospheric argon entrapped during crystallization, and argon loss in weathering. HERZOG, PINSON, and CORMIER (1958) compared the previously published K-Ar dates of WASSERBURG et al., and LIPSON, with Rb-Sr age determinations on glauconites. HURLEY (1958a) determined K-Ar ages of thirteen glauconites of Lower Paleozoic age, finding consistency in results.

In 1959, RUBINSHTEYN, CHIKVAIDZE, KHUTSAIDZE, and GEL'MAN (1959) used glauconite for the determination of the absolute age of sediments. GOLDICH, BAADSGAARD, EDWARDS, and WEAVER (1959) reported on two glauconites from the Upper Cambrian Franconia formation of Minnesota, and illite from shale in the Siyeh limestone of Montana. They concluded that in attempting to date the time of deposition of sediments it is most important to try to understand the geologic history of the sediments subsequent to deposition. HURLEY et al. (1959a), (1960a) compared K-Ar and Rb-Sr ages on a number of Phanerozoic glauconites of known geologic age. They found general concordancy between these two methods, but notices that the observed ages fell below the latest estimates of the true time scale by 10—15 percent. They therefore disagreed with the several investigators who at that time were using glauconite dates for the development of the time scale. The agreement between K-Ar and Rb-Sr age values suggested that the cause of the low ages was due to a process of restructuring of the mineral that affected Ar and Sr equally.

In 1960, the results of extensive Russian argon-dating and retention tests on glauconite were reported by AMIRKHANOV, BRANDT, IVANOV, and TRUZHNIKOV (1960b); AMIRKHANOV and MAGATAEV (1960); KAZAKOV, POLEVAYA, and MURINA (1960); LI, CHEN, TU, TUGARINOV, ZYKOV, STUPNIKOVA, KNORRE, POLEVAYA, and BRANDT (1960); OVCHINNIKOV, SHUR, and PANOVA (1960); POLEVAYA (1960); POLEVAYA, KAZAKOV, and MURINA (1960c) and with SPRINTSSON (1960a, b). POLEVAYA et al. concluded that glauconitic sediments can be dated reliably by the argon method, but that the mobility of K and Ar in different types of glauconite should be investigated systematically. POLEVAYA (1960) presented an absolute geochronological scale derived from glauconite dating in the USSR showing generally good agreement with the Holmes scale. HOLMES (1960a) recalculated Russian age determinations of glauconites to American constants with reference to the age of the base of the Cambrian. FOLINSBEE, BAADSGAARD, and LIPSON (1960) reported on the problems associated with glauconite dating outside of Russia, and suggested the possibility of different orogenic histories in the USSR and elsewhere as an explanation of the higher Russian results. FAUL (1960) noted that evidence was gradually accumulating that the Holmes-Marble time scale should be lengthened, particularly for the Paleozoic, although glauconite ages contradicted this evidence and generally supported the existing scale. EVERNDEN, CURTIS, KISTLER, and OBRADOVICH (1960) investigated argon diffusion in several minerals and concluded that glauconite and illite, due to their fine grain size, are susceptible to high argon loss at 100° C if that temperature is maintained a few million years; knowledge of the burial history of a sample is absolutely essential for each glauconite dated.

After 1960, the testing and use of glauconite added further information on the question of its reliability as an age-dating mineral, and to the sum of points plotted and discussed in the next section. For completeness of the bibliography after 1960 the contributions are listed as follows: EVERNDEN et al. (1961); ALLEN et al. (1964); HURLEY et al. (1961, 1962 a); KLYAROVSKIY et al. (1961); MURINA and SPRINTSSON (1961); POLEVAYA et al. (1961, 1961 a, 1961 b); RUBINSHTEYN et al. (1961); SARDAROV (1963); SCHERBAKOV (1962); VINOGRADOV and TUGARINOV (1961); VOTAKH and DIMITRIYEV (1963); WEBB et al. (1963); WILLIAMS et al. (1962); DODSON et al. (1964); An excellent review of the subject of K-Ar dating of sedimentary and pyroclastic rocks was prepared in 1964 by BAADSGAARD and DODSON for the Symposium on the Phanerozoic time scale (1964). The results of these and earlier investigations are summarized and discussed in the following section.

Comparison of glauconite ages with accepted time scales

It seems that the only way to test a mineral for reliability in age dating is to compare its measured age values with other minerals and other methods. Laboratory tests of diffusion loss of argon do not allow for structural or other readjustments of the mineral to its environment over long periods of time and various depths of burial. In the case of glauconite, as opposed to igneous rock minerals, it has been particularly difficult to find other minerals for direct comparison that occur in sedimentary sequences. As pointed out by FOLINSBEE et al. (1956, 1960, 1961, 1962) biotite from welldated bentonites is considered generally to be the best material for establishing key reference points in the time scale. EVERNDEN et al. (1961) measured several additional well-dated biotites from marine sequences. So far, these biotite K-Ar ages from bentonites appear to fall into quite close agreement with

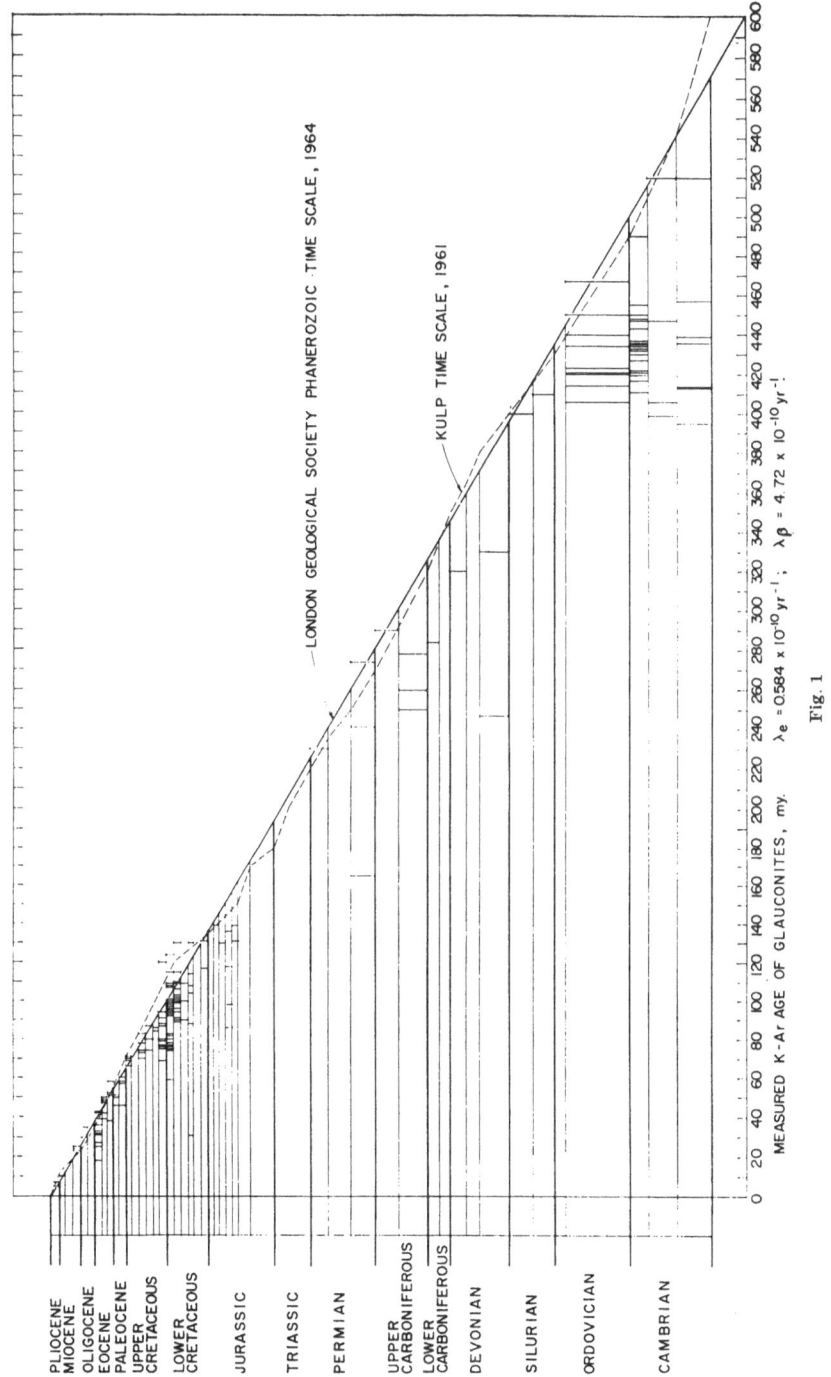

Fig. 1

the best igneous rock time points, and have been used in the construction of the currently accepted time scales.

In order to obtain as much detailed comparison as possible for glauconite K-Ar age values, all of the available published data have been
reduced to the same decay constants and compared with the 1964
Geological Society of London time scale (1964 a). The data are listed
in Table 1 and plotted in Fig. 1. The KULP (1961) scale is similarly
plotted for comparison. It is therefore possible to see at a glance all glauconite measurements relative to the estimated true time scales. This
procedure is similar to, and extends, earlier comparisons by POLEVAYA
et al. (1960), HURLEY et al. (1960a), FOLINSBEE et al. (1960) and several
others who have been mentioned above in the historical review. In most
of these instances the authors concluded that glauconite was usable or
not usable depending upon their own comparison with various time-
scales and the decay constants used.

A study of Table 1 indicates that there is a high degree of concordance
between the various workers and laboratories over the several years of
analytical work. POLEVAYA and KASAKOV and associates have analyzed
glauconites by the volumetric method, using a mass spectrometer to
monitor any possible air contamination. Most of the western analysts
have used isotope dilution techniques. However, the results are obviously
close enough so that the conclusions are not altered by the differences in
laboratory accuracy. Essentially all of the measurements have been
included in Table 1, with the exception of some of the earliest measurements which were omitted because at that time most of the effort was
applied to the methods of analysis, and less to the question of the purity
of the samples.

One of the surprising features of these glauconite measurements is
the rather remarkable consistency of measured ages of glauconite of a
single geologic age from a number of localities over the earth, or the
consistency of a sequence of ages measured on glauconite in a single
geological section. The Upper Cambrian examples demonstrate this.
This has led some workers to the statement that glauconite can be used
as an age indicator with a high degree of reliability. This may turn out
to be the case if certain criteria are discovered by which glauconite can
be selected, but it definitely seems at the present time that an over-all
uncertainty and correction factor must be applied to account for the
discrepancy between the measured glauconite ages and the best values
of the time-scale as indicated by igneous rock minerals.

It is clear from Fig. 1 that the glauconite ages commonly range from
0 to 15 percent low. The age discrepancy appears to be roughly proportional to the geologic age of the mineral suggesting a continuing
effect. Such an effect could be loss of argon, or gain of potassium, or both.

Table 1. *K-Ar Age Measurements on Phanerozoic Glauconites*

Geologic Age of Sediment		Mid-Point Age from L.G.S. scale[1]	Measured K-Ar Age Values on Glauconite[2]	References
Pliocene		4	7, 5	[1], [2]
Miocene	Upper	10	10	[2]
	Middle	15		
	Lower	22	23, 25, 24	[2], [3], [4], [5]
Oligocene	Upper	28		
	Middle	31	30	[6]
	Lower	35		
Eocene	Upper	41	43, 31, 43, 33, 25	[2], [3]
			18, 39, 36, 27	[2]
	Middle	47	48, 50, 42, 47, 39	[2], [3]
	Lower	51	58, 52, 51, 38	[1], [2]
Paleocene	Upper	56	50, 46, 50	[2], [3], [7]
	Middle	60	60, 57, 46, 58	[1], [2]
	Lower	63		
U. Cretaceous	Maestrichtian	67	66, 70, 70, 70	[2], [3], [7]
	Campanian	73		
	Santonian	79	83, 70, 73, 74	[2], [3], [6]
	Coniacian	85	87, 80, 74	[2], [3], [6]
	Furonian	91	84, 86, 90, 88	[1], [2]
	Cenomanian	97	79, 91, 88, 76, 75, 94, 87	[1], [2], [3]
			79, 69, 121	[8]
L. Cretaceous	Albian	103	75, 78, 109, 115, 82, 108, 99	[9]
			59, 82, 75, 75, 108	[10]
			96, 97, 94, 94	[2]
			98, 97, 124	[11]
			97, 80, 95, 78	[3], [6], [12]
	Aptian	109	90, 90, 100	[3]
			90, 130	[6]
			102, 110, 115	[2]
			94, 101, 103, 95, 101	[11],][12], [13]
	Barremian	115	100, 109, 90	[3], [6], [11]
	Hauterivian	118	88, 104, 108, 114, 31, 130	[1], [2], [3], [11]
	Valanginian	127		
	Ryazanian	133	131, 117	[11]
U. Jurassic	Purbeckian	138		
	Portlandian	143	119, 132, 129, 107, 129, 117	[11]
			139, 125	[2]
	Kimmeridgian	148	130, 135	[5], [6]
	Oxfordian	154	86, 98, 118, 136	[2], [10]
	Callovian	159	139, 131	[2]
Jurassic	Middle	167		
Jurassic	Lower	182		
Triassic		210		
Permian	Upper	235	230	[5]
	Middle	250		
	Lower	270	241, 274, 165	[1], [5]
Pennsylvanian	Upper	290	290	[5]
	Lower	315	250, 260, 278	[1]
Mississippian	Upper	330	284	[1]
	Lower	340		
Devonian	Upper	352	320	[5]
	Middle	365		
	Lower	382	330	[1]

Table 1 (continued)

Geologic Age of Sediment		Mid-Point Age from L.G.S. scale[1]	Measured K-Ar Age Values on Glauconite[2]	References
Silurian	Upper	405	400	[6]
	Lower	425	410	[1]
Ordovician	Upper	440		
	Lower	475	434, 420, 406, 423	[2]
			440, 450, 414	[1], [3]
Cambrian	Upper	510	421, 435, 432, 433, 420	[1]
			434, 427, 417, 411, 422	[1]
			435, 436, 448, 455, 447	[1]
			430, 450, 443, 490	[5], [7], [14]
	Middle	530	406, 447, 399, 520	[2], [5]
	Lower	555	413, 457, 436	[2]
			413, 395, 439, 520	[1], [5], [10]

[1] London Geological Society 1964 Phanerozoic Time Scale gives the age of the base of the stratigraphic unit. The Mid-Point values are taken by interpolation. [The Phanerozoic Time-Scale, ed. by W. B. HARLAND, A. G. SMITH, and B.WILCOCK. Quart J. Geol. Soc. London, 120 S, 0—458 (1964)].

[2] Reduced to same value of K^{40} decay constants: $\lambda_e = .584 \times 10^{-10} yr^{-1}$; $\lambda_\beta = 4.72 \times 10^{-10} yr^{-1}$.

References for Table 1

[1] HURLEY et al. (1960a).
[2] EVERNDEN et al. (1961).
[3] POLEVAYA et al. (1961).
[4] POLEVAYA (1961b).
[5] POLEVAYA et al. (1960c).
[6] AMIRKHANOV and MAGATAEV (1960).
[7] WASSERBURG et al. (1956).
[8] RUBINSHTEYN et al. (1959).
[9] WILLIAMS et al. (1962).
[10] FOLINSBEE et al. (1960).
[11] DODSON et al. (1964).
[12] AMIRKHANOV et al. (1958b).
[13] RUBINSHTEYN (1961).
[14] GOLDICH et al. (1959).

In the chapter on argon leakage there is much discussion on the subject, which will not be repeated here. It should be noted, however, that glauconite (and illite-montmorillonite) may be special cases in which the first-formed mineral is not well-ordered, and in which there may be a process of continuous readjustment during burial. This is implied in the statements of several investigators of argon loss (eg. AMIRKHANOV and associates) who have noted different modes of release, and have considered these to be related to subcrystalline phase changes or progressive restructuring, with the argon partly trapped in the phase boundaries.

HOWER (1960) has attempted to explain the discrepancy as follows. He has noted that there is an increasing amount of expandable layers

in glauconites that are increasingly less pure, and that an inverse relationship exists between percent expandable layers and amount of potassium. Glauconites with 30% expandable layers contain about 3%
potassium whereas the purest glauconite with about 10% expandable
layers contains about 6% potassium. The pure 1 Md contains more than
7% potassium. HOWER also concluded that glauconites with a small
amount of expandable layers (10% or less) commonly occur dominantly
in clean sandstones and limestones, and he also notes appears to be a
significant difference in glauconite structure and composition with
geologic age. The discussion on the problem of illite given below is rather
similar, and adds more uncertainty to the whole question of loss of
argon, gain of potassium or partial restructuring of the mineral. It is
clear that simple laboratory determinations of diffusivity in an environment that is not similar to the deeply buried sedimentary environment
must be interpreted with caution.

The use of glauconite in dating the Proterozoic

Although glauconites may not show the true age, they seldom show
an age that is more than 10 or 15 percent younger than the true age of
sedimentation, and may be useful as an indicator of the approximate
true age if this is allowed for. Also glauconite may serve to correlate
formations in the late Precambrian if the stratigraphic sections are
relatively similar in their depth and history of deformation. Some useful
correlations of this kind have been made.

POLEVAYA, KAZAKOV and associates have reported on Precambrian
glauconite age measurements, with results summarized in Table 2.
If these glauconite measurements are corrected upward slightly, they
form a basis for a tentative Proterozoic time-scale for the central parts
of Europe and Asia.

LI and associates (1960) have carried out a few measurements on the
Sinian of the Chinese-Korean platform, with results on glauconites
that range from 800 to 1040 m.y., also shown in Table 2. This suggests
the possibility of correlating various sequences in the Sinian with the
general stratigraphy of the Ripean of the Russian platform. VOTAKH
and DIMITRIYEV (1963) have measured Precambrian formations in the
Igarka and Turukhansk areas, so that this region may also be tentatively
tied in to the time-scale.

In North America there has been very little direct measurement of
authigenic minerals in the Proterozoic. HURLEY et al. (1962a) determined
the K-Ar and Rb-Sr age of mixed-layer illite-montmorillonite in the
Gunflint formation in the north limb of the Huronian geosyncline.
The measured age of 1600 m.y. was considered a minimum age, and the
estimated true age was given as 1900 ± 200 m.y.

Table 2. *K-Ar Age Measurements on Proterozoic Glauconites*
[Polevaya et al. (1961)]

Region	Series	Suite	Age Values, m.y[1]
Bashkir Antichlinorium, classified as Rhiphean		Ashinskaya (may be L. Cambrian)	572
	Valdai	Laminarite	598, 606
	Karatauskaya	Ukskaya	618, 616
		Minyarskaya	760
		Inzerskaya	932, 876
		Katarskaya	
		Zilmerdakskaya	
	Yurmatinskaya Burzyanskaya	Avzyanskaya	1263
Ryazan'-Pachelma Depression	Serdobsk	Pachelma	767, 770
		Pachelma	726, 743
		Lower Pachelma	898, 943
		Lower Bavlinskaya	1290
Murmansk-Kola Region Believed equivalent of Serdobsk and Sinian			865, 1040 1059, 904 1018
Siberia, Aldan shield		Motskaya	609
		Chividenskaya	747
		Pogoryuyskaya	1140
		Solooliyskaya	1263, 1260
Chinese Platform, Hopeh Province	Sinian, 10th Div.	Chinerhyu horizon	873, 890
	Sinian, 4th Div.	Chinerhyu horizon	1040

[1] Polevaya et al. used decay constant: $\lambda_e = .557 \times 10^{-10} \mathrm{yr}^{-1}$; $\lambda_\beta = 4.72 \times 10^{-10} \mathrm{yr}^{-1}$.

In Australia, Webb et al. (1963) have dated three Proterozoic glauconites. Two of the samples from the Aquarium formation and one from the Rosie Creek sandstone, Tawallah group, Carpentaria region, Northern Territory, gave K-Ar ages of 1475, 1580, and 1525 m.y. respectively. These ages conform with the maximum and minimum possible ages for the Tawallah group as given by the underlying Norris granite (1780 m.y.). Isotopic analysis of the lead from a syngenetic galena found in the overlying conformable McArthur group gives a model age of about 1560 m.y. The authors conclude that there is almost complete retention of radiogenic argon in these samples.

Evidence from glauconite on the age of the base of the Cambrian

The base of the Cambrian has been a matter of controversy for many years. Glaessner (1963) has recently reviewed the discussion on the subject by Polevaya and Kazakov (1961), Kulp (1961), Wetherill (1960), Holmes (1960a) and others. He contends that the problem of the base of the Cambrian is one of biostratigraphy and correlation with the type area for the Cambrian System in Wales. This type section cannot be used to define its lower boundary because the basal part is absent. However, by selecting a subsidiary type area for the basal section, and a continuing sequence below it with a minimum break, GLAESSNER chooses to place the boundary at the base of the oldest known rock containing Cambrian animal macrofossils. For this purpose he selects the Cambrian section in Scandanavia and in the Baltic Region which adjoins the Caledonian trough. Here the Blue Clay of Estonia and Leningrad contains the oldest known Cambrian fossils, and further east it is underlain by similar rocks without such fossils. He extends the stratotype of the Cambrian System to include as its basal unit, the Blue Clay and its equivalents (Baltic Series). The further extension into rocks that are similar but without Cambrian fossils could go on indefinitely and with increasing confusion, so that he justifiably chooses this criterion for the definition of the beginning of the Phanerozoic.

The dating of the Baltic Clay and its equivalents elsewhere is therefore important. Fortunately, it contains glauconite. K-Ar dates by POLEVAYA and KAZAKOV (1961) average 540 m.y., or 520 m.y. if corrected to the constants used in this review. The Blue Clay is underlain by a sandstone formation known as the Super-Laminarites Beds, constituting the lower part of the Baltic Series. In the northern part of the Russian Platform the Baltic Series is underlain by the Laminarites Clay and the Gdov Sandstone making up the Valdai Series, which has yielded glauconite ages averaging 560 m.y. by POLEVAYA and KAZAKOV, when reduced to the same constants. GLAESSNER has good grounds for not accepting the arguments for placing the Valdai Series in the Cambrian. This would place the base of the Cambrian between the glauconite ages of 520 and 560 m.y., in reference to such ages in the sediments of the Russian Platform.

It is quite likely from the discussion of Fig. 1 that these glauconites have lost argon and/or gained potassium. Allowing for this and using other data, GLAESSNER estimates the base of the Cambrian to be between 550 and 570 m.y.

Measurements on whole-rock samples of shale or its mineral components

In 1959 HURLEY et al. (1959 b) discussed the question of authigenic versus detrital illite in sediments, referring to K-Ar studies on the clay-size fractions in materials deposited recently. In 1961 they reported on K-Ar measurements on clays in recent sediments from the Mississippi and other rivers, and illite in clay soils. Selected results are given in

Table 3. *K-Ar Age Measurements on the Clay Size Fractions from Pelitic Sediments*

Material	Geologic Age	% K	K-Ar Age Value, m.y.	Reference
Mississippi River Delta	Recent	2.2	102	HURLEY et al. (1961a)
		2.1	175	
		2.6	206	
		2.3	166	
		2.5	160	
		2.5	145	
Ditto, Acid treated		2.4	178	HURLEY et al. (1961a)
		2.3	160	
		2.3	206	
Red River sediments	Recent	3.1	483	HURLEY et al. (1961a)
		2.5	830	
Rappahannock River bottom sediments	Recent	1.5	334	HURLEY et al. (1961a)
		1.5	290	
		1.5	460	
		2.3	497	
Venezuelan Shales from drill cores	Upper Miocene	2.1	115	HURLEY et al. (1963a)
		2.2	122	
		2.2	127	
	Lower Miocene	2.4	116	
		1.9	210	
		2.1	121	
		2.1	202	
Shale outcrops, Texas	Mid-Cretaceous	2.4	170	HURLEY et al. (1963a)
		2.5	178	
		3.0	175	
Shale	U. Pennsylvanian	5.1	246	EVERNDEN et al. (1961)
	Westphalian	2.9	295	
	L. Mississippian	2.1	305	
Shale, Wyoming Co. N.Y.	U. Devonian	4.6	334	HURLEY et al. (1963a)
Shale	L. Devonian	4.2	347	EVERNDEN et al. (1961)
	Basal Devonian	5.3	362	
Shale, Ontario Co, N.Y.	U. Silurian	1.4	326	HURLEY et al. (1963a)
Geneseo Co. N.Y.	U. Silurian	4.7	332	
Rochester N.Y.	M. Silurian	5.4	363	
Rochester N.Y.	M. Silurian	5.6	357	
Sodus fm. N.Y.	M. Silurian	6.1	367	
Shale	L. Silurian	5.6	284	EVERNDEN et al. (1961)
	L. Silurian	5.7	267	
Shale Rochester N.Y.	U. Ordovician	6.5	374	HURLEY et al. (1963a)
Shale Rochester N.Y.	U. Ordovician	5.7	368	
Shale	M. Ordovician	2.8	485	EVERNDEN et al. (1961)
	U. Cambrian	5.8	457	
	M. Cambrian	6.1	645	
	M. Cambrian	6.2	708	
	M. Cambrian	6.0	646	
	M. Cambrian	5.9	530	

Table 3. The clay-size fractions all show ancient K-Ar ages. This is compatible with the observation by Hurley et al. (1961a) that the illite from the soil above a Devonian shale showed no decrease in age below the value for the illite in the fresh shale below, despite a decrease in K content. It is therefore clear that illite may be recycled, and 1 M material may be deposited as ancient detritus.

In 1961, Evernden et al. (1961) presented a list of K-Ar ages on illite-bearing sediments. These are included in Table 3. It can be seen that the illite ages may fall below or above the time scale, by comparison with the data in Table 1. Hurley et al. (1963a) measured more samples of clay-size fractions from shales of different geologic ages with results also included in Table 3.

In a test of the different components of a shale, Bailey et al. (1962) separated the illite in a Pennsylvanian shale into 2 M_1 and 1 Md poly-types. The former gave a K-Ar age considerably older than the Penn-sylvanian and the latter an age less than half the age of the Pennsyl-vanian. The authors believed the low age in the 1 Md component to be due either to preferential Ar loss because of the small particle sizes, or to reorganization and K-fixation in montmorillonite and degraded micas in post-Pennsylvanian time. The data are included in Table 4.

This investigation was followed by a more detailed size fractionation of a shale by Hower et al. (1963). The Upper Ordovician Sylvan shale was separated into 6 size fractions ranging from 0.08 to 62 microns.

Table 4. *Comparison of K-Ar Ages on Different Size Fractions from Pelitic Sediments: Mostly Illite in Fine Sizes*

Material	Geologic Age	Size Fraction	% K	Age Value	Reference
Mississippi R. Delta	Recent	-2μ	2.2	102	Hurley, et al. (1961a)
		$+2\mu-400$ mesh	2.2	256	
		$+400$ mesh	1.5	405	
1 Md Concentrate	Pennsylvanian	-0.2μ	3.5	117	Bailey et al. (1962)
from shale, Iowa		-0.2μ	3.3	133	
		-0.2μ	4.5	210	
		-0.2μ	4.2	114	
		-0.2μ	4.7	233	
2 M_1 Concentrate	Pennsylvanian	$2.0-0.2\mu$	3.1	348	
from shale, Iowa			4.1	310	
			3.6	385	
Sylvan Shale, 1 Md	U. Ordovician	-0.08μ	5.1	360	Hower et al. (1963)
1 Md		$0.08-0.2\mu$	5.4	350	
1 Md		$0.2-0.45$	5.4	428	
1 Md $+$ 2 M		$0.45-2$	4.2	462	
2 M $+$ 1 Md		$2-6$	3.2	510	
2 M $>$ 1 Md		$6-62$	2.0	540	
whole shale			3.8	450	

The whole rock gave an age value of 450 m.y., which is in approximate agreement with the estimated true age of sedimentation. However, this agreement was fortuitous, as the different size fractions showed a spread in age values from 350 m.y. for the finest to 540 m.y. for the coarsest, as given in Table 4. X-ray analysis of the materials in each size fraction indicated 1 Md illite in the finer fraction and 2 M muscovite in the coarser.

It appears from these studies that there is no zero-age, fine-size, illitic material deposited in sediments at the outset, but that the detrital age is gradually wiped out during burial and argon starts to accumulate at the same time in reordered material. The aggregate age of the clay-size fraction gradually decreases from an excess age in young sedimentary rocks to a deficiency in age that is commonly about 15 percent too young in older sedimentary rocks. Superimposed on this is a resistant detritus that is made up of 2 M muscovite (and probably other K-bearing igneous minerals), which adds an excess age value to the shale aggregate as a whole. Frequently, the sum of the two components, older 2 M muscovite and younger and 1 Md illite, is nearly balanced out and the apparent K-Ar age on the total shale turns out to be quite close to the time age of the sedimentation. However, this is fortuitous and not to be relied upon.

With these observations in mind, the conclusions of NOVIKOV (1963), and PERMYAKOV and SAVCHENKO (1963) are understandable. These investigators found that the apparent K-Ar of shales appeared to agree closely with a true biostratigraphic age in the cases which they analyzed.

Other possible authigenic materials have been tested to determine whether they retain radiogenic argon. The problem is generally made more difficult by very low concentrations of potassium. LIPPOLT and GENTNER (1962) tested some limestones and fluorites, and samples of whole bone, bone apatite, and tooth of Triassic age. The results were generally low and spotty, with one limestone sample and the bone sample yielding a K-Ar age approximately in line with the true age. The hydrothermal fluorites showed excess argon in general. Another attempt was made by CHERDYNTSEV and KOLESNIKOV (1964) who investigated carbonate rocks of sedimentary origin. They reported that carbonate materials retained their radiogenic argon well in tectonically quiet areas, but lost it almost completely in tectonically active areas. In another report, LIPPOLT and GENTNER (1962) tested various fossil materials with generally low results. A sample of biotite believed to have been authigenic, from the Utica shale of Upper Ordovician age near l'Epiphanie, Quebec was measured by BEALL (1962), who found a K-Ar age of 402 m.y. and Rb-Sr age of 550 m.y. So far, therefore, it appears that attempts on authigenic minerals other than glauconite show little hope for sediment dating. This does not include minerals such as biotite in bentonite or other volcanic materials, which are considered in another chapter.

K-Ar age measurements on pelagic sediments

Hurley (1958b) reported a K-Ar age of 80 m.y. on a sample of bottom sediment from the Pacific Ocean, and suggested that the K-bearing material was largely detrital in origin. In 1960 he reported K-Ar ages on a number of pelagic red clays from deep oceanic areas (Hurley, 1960). In 1963, he listed with others (Hurley et al., 1963b) the findings that are summarized in Table 5. The general conclusion was that pelagic red clays in the North Atlantic contain potassium-bearing phases that are dominated by detrital illite, much of which may have been derived from the weathering of Paleozoic and younger shales on the continental areas. The material was estimated to have been largely airborne in the deep oceanic areas, but in the Caribbean Sea appeared more likely to have

Table 5. *K-Ar Age Measurements on Pelagic Red Clays and Foraminiferal Sediments*

Location	Material	Depth, m.	% K	Age Value, m.y.	Reference
Indian Ocean					
South of Cocos Islands	Red Clay	5501	2.1	250	Krilov et al. (1961)
Latitude of Java	Red Clay	4853	2.1	80	
South Africa	Foram. Sed.	1808	1.8	190	
South Africa	Foram. Sed.	4400	1.6	250	
Sumatra area	Foram. Sed.	4512	1.0	155	
Pacific Ocean					
N.W. part of ocean	Red Clay	5520	2.0	155	Hurley (1958b)
N.W. part of ocean	Red Clay	5000	2.4	145	
20° N; 125° W	Red Clay	5450	2.7	80	
Atlantic Ocean					
27° N; 61° W	Red Clay[1]	5900	2.3	210	Hurley et al. (1963b)
28° N; 61° W	Red Clay	5300	3.0	85	
27° N; 34° W	Red Clay	550	2.5	252	
Mid-Atlantic Ridge Rise					
off Sahara	Red Clay	—	2.0	219	Hurley et al. (1963b)
24° N; 64° W	Untreated Red Clay	5949	2.9	417	
24° N; 64° W	−2 μ fraction		2.2	464	
24° N; 64° W	+2 μ fraction		2.7	390	
00° S; 24° W	100 cm down core[1]	2430	1.7	240	
00° S; 24° W	200 cm down core	2430	1.7	323	
00° S; 24° W	350 cm down core	2430	0.9	233	
00° S; 24° W	400 cm down core	2430	1.9	163	
Caribbean Sea					
15° N; 68° W	170 cm down core	4180	2.0	225	
15° N; 68° W	270 cm down core		1.8	226	
15° N; 68° W	350 cm down core		1.9	(225)	
15° N; 68° W	515 cm down core		2.0	222	
15° N; 68° W	−2 μ fraction only				
	170 cm down core		1.9	163	
	270 cm down core		2.5	168	
	515 cm down core		2.1	153	

[1] Sample treated with cold acetic acid to remove carbonate.

been waterborne and carried to the sea by rivers. The question of transport however, is not relevant to the fact that the K-bearing phases appear almost entirely to have been closed mineral systems prior to their deposition in the deep oceanic areas.

In 1961, KRILOV, LISITSYN, and SILIN (1961) also published an interesting investigation of the K-Ar ages in the components of oceanic sediments. A summary of their measurements is given in Table 5. In general, they concluded that radiogenic argon is preserved fully or nearly so in the finest (up to less than .001 millimeters) particles of feldspar and mica, which they found to be the components of most clay; and also that there was no appreciable absorption of potassium from oceanic water. Earlier, KRILOV and others had studied the preservation of radiogenic argon in the products of disintegration of rocks to clays, and had concluded that the K-Ar relationship was preserved even in the disintegration products. This led to the belief that most oceanic oozes can be used for the determination of the absolute age of the provenance from which the detrital material came. KRILOV and his associates found that the fine fraction of deep sea clays contained up to 5 or 10 percent microcline and as much as 10 percent of biotite. They therefore believed that these detrital minerals dominated the K-Ar age in the sediment.

The age of detrital components in clastic sediments in the study of paleotectonics

The first serious attempt to use the K-Ar method for interpretation of paleogeography appears to have been carried out by VISTELIUS (1959), who studied the origin of the Cretaceous red beds of the Cheleken Peninsula. The work was extended (VISTELIUS and KRILOV, 1961) to a study of the arenaceous siltstones in southwest central Asia, east of the Caspian Sea. In a later report VISTELIUS in 1964 classified the mid-Cretaceous sandstones into three groups having dispersion centers of late Precambrian and Triassic ages, and an area of intrusion dated at 270 m.y.

KRILOV and various associates have also measured the K-Ar ages on a number of clastic materials, and correlated these with source areas (KRILOV, 1961; KRILOV and SILIN, 1959, 1960). In 1960, KRILOV and SILIN reported on a study of the argon age in the components of Archeozoic sandstones and clastic clays, and showed that in most cases the original ratios had been preserved. In 1959 the same authors traced the provenance of the terrestrial sediments in the Baltic glacial deposits and the Black Sea coastal sands.

DUBAR' (1962) made a similar study in the basin of the Lena, with both K-Ar and Rb-Sr analyses carried out in GERLING's laboratory.

Biotite from the Mesozoic sediments of the Aldan and Stanovoy areas was found to correlate in age with the Precambrian complexes in the Verkhoyansk region to the east. In general, it is clear that the K-Ar method will be extremely useful in the correlation of sedimentary materials, a study of their provenances, migration-directions, and distances of travel. It is particularly noteworthy that the effect of weathering does not seem to be great. Mineral separations will probably be necessary to assure that age measurements are meaningful and not mixtures of authigenic or older materials of different age.

The Problem of Contamination in Obtaining Accurate Dates of Young Geologic Rocks

By

G. H. CURTIS

Introduction

There now exist at least two methods of radiometric dating covering most parts of the geologic time-scale from essentially 0 to 4.5×10^9 years, which fact puts geochronology on a reasonably sound basis. Almost all systems of natural radioactive decay are now being used to obtain more geologic information than just radiometric dates, although radiometric dates in themselves throw light on a great number of geologic problems. The K/Ar system of dating does not have the diversity of geologic applications the Rb/Sr and U/Pb methods have owing to the chemical inertness of argon, but it is this very property of argon coupled with its large size and the ubiquity of its parent, potassium, which has given the method an almost unique position in dating geologic events with high precision in the range of 50,000 to 50,000,000 years, here termed "geologically young". Older rocks, of course, can be dated by K/Ar but other decay schemes may be applied equally well. New investigations of the U/He method, of intermediate decay members of the two uranium series, and fission track counting give promise of ending the unique position of the K/Ar method in dating young rocks, but each of these schemes so far has had either limitations of application more severe than obtain for the K/Ar method or has lacked the precision obtainable with the K/A scheme.

Many questions remaining to be answered in geology are concerned with rates and with the duration of various types of physical and chemical conditions which have obtained on and below the surface of the earth. For instance, concerning the geologic time-scale itself and its possible degree of refinement; what are the rates of evolution of the various life-forms and what are the rates of dispersal of faunas and floras? In short, what is the degree of synchroneity of the end of a geologic era, a geologic period or a geologic epoch on the different continents? The answers to these questions are of great importance to the fields of paleontology and anthropology as well. In the seemingly everflexing, deforming and shifting crust of the earth, what are the rates of subsidence in the various types of geosynclines and how constant are they? What are the rates of deformation during orogenies and what are the rates of erosion of the land-forms produced? If climate is related to land-forms and continental emergence and submergence, then this question bears on the cause of the glacial advances and retreats of the Pleistocene epoch so that in this respect the question of when the continental glaciations began in the Pleistocene becomes important as well as what the durations were of each of the separate glacial advances. As VERHOOGEN (1960) has pointed out, crustal activities are related to heat-flow in the earth, of which volcanic activity is one manifestation, but how long do magma chambers persist beneath volcanoes and how long does it take to fractionate a magma into the contrasting types seen at igneous centers? How long do ore-depositing fluids persist during and after igneous activity?

Questions such as these are being tackled in a number of radiometric laboratories today, and preliminary answers are beginning to be published, but much remains to be done. For the most part, these questions lie beyond the scope of ^{14}C and can only be undertaken by methods which can resolve accurately intervals of a few tens of thousands of years during the past one million years and intervals of a few hundred thousand years during the past 50 million years. So far, the K/Ar method is the only one proven to have this resolving power.

While dates obtained by all radiometric methods for older geologic rocks are suspect because of the increasing probability with age of their having been subjected to metamorphic conditions with consequent diffusion losses of the radioactive and radiogenic isotopes, dates obtained for younger rocks are suspect because of the increasing importance of contamination with decreasing age. For example, one tenth of one percent of contamination of billion year old material in a hundred million year old sample increases the age of the sample by only one percent or one million years, an insignificant amount which will be hidden in the probable error; but this same contamination in a sample one hundred thousand years old increases the age one thousand percent!

Because of the importance of eliminating contamination from young samples in order to get accurate dates and because contamination can so easily go unrecognized, this chapter is devoted to a discussion of the causes of contamination and to criteria for its recognition.

Contamination

All substances in the sample to be dated radiometrically — minerals, rock fragments or glass, adhering colloidal particles or adsorbed gases-which in most cases were not formed at the same time as the sample itself, hence which may give an erroneous age to the sample, are here called contaminants. Contamination can occur during formation of the sample or later. I shall first discuss contaminants added to the sample at the time of its formation.

Material used for K/Ar dating is presumed to have crystallized or cooled (volcanic glass) at the time of the event to be dated. Potassium-bearing crystals have been used for dating that have grown at temperatures of only a few degrees centigrade – authigenic minerals such as the clays and some feldspars, and various salts – up to a thousand degrees or so centigrade in the case of the igneous minerals. During their growth it is essential that potassium-bearing crystals do not incorporate radiogenic argon present in their enviroment in the form of dissolved gas or in older crystals. Authigenic minerals growing in near-surface environments at low to moderate temperatures are particularly likely to form as out-growths on older minerals of the same type or in rocks containing the same minerals. Volcanic tuff deposits are noted for their authigenic minerals, but usually in tuffs it is the primary minerals that are of interest for age determination rather than the authigenic minerals which only serve to decrease the obtained radiometric age. In some cases, however, (Hay, 1963) the investigator wants the age of the authigenic minerals, and criteria must be found for distinguishing them from the primary. Differences in crystal shape and colour are the usual means for such distinction, and the sample is prepared by hand-picking under a bi-nocular microscope. When the mineral grains are very fine, as are usually the clays, it becomes virtually impossible to distinguish authigenic from primary, and if there is a suspicion of primary clay being present in the sample, the situation is hopeless. Glauconite is probably the most commonly used authigenic mineral, and although contamination from older glauconite brought into the environment of glauconite formation from eroded older beds seems potentially a serious threat, there have been very few cases where such contamination has been suspected. More frequently glauconite may incorporate older detrital grains of feldspar and other potassium-bearing minerals. These contaminating grain-fragments may be easily detected in thin-sections of grain-mounts or by crushing the glauconite pellets in a mortar and examining some of the mixture in a drop of refractive index oil with a polarizing microscope.

Owing to the high amount of air adsorbed or incorporated in authi-genic minerals of all types, most of them cannot be used for obtaining accurate radiometric dates younger than approximately one million years, for when the air argon in a sample rises much above 90 percent of the total argon extracted, the precision of the argon analysis falls off rapidly and with it the accuracy of the age determination. This is easily understood from a consideration of the precision obtainable in measuring the $^{36}Ar/^{40}Ar$ ratio, the ratio needed to correct for the presence of air argon in the spike and sample argon mixture. This ratio can be measured with a precision of approximately ± 1 percent. If E = the percentage error in radiogenic ^{40}Ar, e = the percentage error in the $^{36}Ar/^{40}Ar$ ratio, and r = the percentage of radiogenic argon in the extraction, we can write

$$E = \frac{e(100\text{-}r)}{r}$$

From this it may be seen that as r approaches 1 percent, E approaches 99 percent. Now, the air argon in most authigenic minerals is about 95 percent of the total at 1 million years of age and is, of course, higher at younger ages. Because this air-argon occurs throughout the mineral, having been incorporated during growth, it cannot be removed by heating or acid-treatment without liberating the radiogenic argon also. As there is very little dissolved air in magma, the problem of air-argon contamination does not occur in igneous minerals through incorporation into the crystal lattices during growth but by adsorption onto the surface of the crystals after they come into contact with air at a later date. In so far as this adsorbed argon can be more or less removed by acid treatment and heating (EVERNDEN and CURTIS, 1965), the mineral can be dated at much younger ages than can be done with authigenic minerals; however, ultimately the amount of residual air-argon approaches 100 percent under the most favorable treatment and the precision consequently falls off severely. This limit is generally reached with the best material – sanidine and leucite – at between 5,000 and 10,000 years.

The question of how much initial radiogenic ^{40}Ar may occur in igneous minerals as a result of incorporation during growth in a magma containing some radiogenic argon has not been fully resolved. In three diverse types of lava of recent age – Ischia rhyolite flow of 1304 a.d., Katmai rhyolite pumice of the 1912 eruption, and trachyte from an eruption in New Guinea in 1954 – we found the contents of ^{40}Ar to be less than would give an age of 10,000 years, or within the experimental limits of detection (unpub. data, this lab.). To these essentially zero ages may be added the dates we have obtained on rocks that appear from geological considerations to be young: 30,000 years for pumice at the mouth of Lake Naivasha, Kenya (EVERNDEN and CURTIS, 1965); 60,000 years for the oldest and 5,000 years for the youngest of the Mono Craters, California (ibid.): 30,000 years for the black tuff from Sorrento, Bay of

Naples, Italy (unpublished data, this laboratory). This black tuff has been dated at 35,000 years by ^{14}C at the University of Rome according to SEGRÉ (pers. comm.). On the other hand, GENTNER and LIPPOLT (1963) give evidence of the presence of excess initial ^{40}Ar in several tuffs of the Eifel volcanic district of Germany, one of which yielded an age at least 2 million years greater than it should be based on ages of tuffs below it. They pointed out, however, that the presence of contaminating grains of older material could not be definitely ruled out. With respect to these anomalous dates of GENTNER and LIPPOLT, it should be mentioned that professor FRECHEN at Bonn gave us a sample of sanidine collected in the Eifel district for which an Allerod age based on ^{14}C work was indicated, i.e., approximately 12,000 years. We obtained 5 different ages from splits of this sample which varied from 200,000 to over 700,000 years. Clearly, these discrepant ages were not caused by inherited argon which would have been evenly distributed throughout the grains: small amounts of undetectable contamination – probably sanidine – of very ancient age which was not evenly distributed throughout the 5 splits seems the most probable explanation.

On theoretical grounds, one might expect to find many more cases of the presence of noticeable amounts of initial ^{40}Ar in igneous rocks than so far have been detected. Magmas formed at depths of 50 to 100 kilometers are under sufficient confining pressure to keep significant quantities of old radiogenic argon in solution, argon that has been formed from ^{40}K decay deep within the earth. Crystals growing in this environment should incorporate some of this argon into their lattices even though argon is inert. While we might anticipate that lava extruded slowly at high temperatures would outgas completely before cooling, many gas rich magmas are erupted into the atmosphere with sufficient rapidity to cause them to explode upon the sudden release of confining pressure. These explosions often result in more or less fragmenting the magma into fine particles of glass and crystals which upon rising quickly high into the atmosphere cool rapidly. Such crystals and glass shards, the principal constituents of tuffs, might sometimes be expected to have cooled before being thoroughly outgassed, but so far, as pointed out, very few cases of this have been documented. To those cases from the Eifel mentioned by GENTNER and LIPPOLT should be added the following, though again not definitely proven, one. Duplicate runs of glass and sanidine samples concentrated from a single tuff in western Nevada by POWERS of the U.S Geological Survey gave ages of 12 and 2 million years respectively. Had the dates been reversed and the sanidine fraction been older, either contamination from older potassium feldspar or diffusion loss from the glass could be suspected, but these explanations appear to be ruled out. A count of over 1,000 glass shards in the sample showed no mineral-grain contamination.

From all the work done at Berkeley it appears that contamination from old potassium-bearing minerals is a far more likely source of excess radiogenic argon in samples than argon still left in the magma at the time of eruption, such older potassium-bearing minerals either being torn from the side of the volcanic conduit during eruptions or added to the erupted constituents by contemporaneous erosional processes. Sometimes their presence is difficult to detect by ordinary optical examination, as in the case of the Eifel sample of FRECHEN.

Foreign mineral contamination of tuff samples being so important to the problem of obtaining accurate young geologic dates, it appears obvious that criteria for its recognition should be discussed, but since the various modes of formation of fragmental volcanic debris composing tuffs bear significantly on the type and quantity of contamination a tuff may be expected to have, these modes of formation must be discussed first as an essential part of the problem.

The concept of gas-charged magma rising up a conduit from great depth in the earth and vesiculating as the confining pressure decreases is easily envisaged, together with its consequent violent discharge as siliceous froth onto the surface of the earth or into the atmosphere as a plasma of gas and particles or crystals and glass. Deposits of ash so formed are generally readily recognized by the curved shapes of the glass shards formed from the explosion of bubbles and the presence of pumice fragments. The associated primary igneous crystals are usually easily identifiable by the glass films still clinging to or enveloping them. Most tuffs interbedded with clastic sediments of erosional origin and occurring far from centers of volcanism are probably of this type. Such tuffs must obviously have been borne by the wind to their sites of depositon. Closer to volcanic centers, however, finely comminuted debris of igneous types closely resembling ash as described above may have been fragmented by a variety of other processes. There is an extensive literature about these processes, and a good review of them is given by FISHER (1960). For our purposes here, we need mention only that, depending on the rate of ascent of the magma in its conduit, the amount of loss of heat and volatiles from the rising magma, the degree of crystallization of the magma and the viscosity of the magma, the magma may partly solidify then fragment by friction before reaching the surface or shortly thereafter. The term friction as used here is interpreted broadly to include, in addition to abrasion of the magma with its walls, fragmentation of viscous cool magma in a rising column or in an extrusive lava flow by more mobile fluid magma pushing from behind. To be sure, frictional processes may be aided by vesiculation in greater or less degree.

Fragmentation by these less explosive processes produces in some cases huge quantities of more or less finely comminuted igneous debris

which are not ejected into the atmosphere to be distributed by the winds
but are debouched into stream tributaries on the earth's surface in the
vicinity of the conduit. While some explosive action may accompany this
type of activity and resulting clouds of ash may be carried by winds,
explosive action appears to be minor. Gravity and water act on these
masses of debris and carry them in the form of volcanic mud-flows toward
the nearest basin of deposition. Some of the finest material may be sorted
from mudflows during transport or by subsequent erosion after deposition
and carried long distances from the source to be deposited in beds which
look superficially like wind-born ash deposits of explosive origin. The
virtual absence of curved glass shards together with the high proportion
of crystals such deposits contain are important criteria for their recognition.

These two major types of igneous processes causing fragmentation
of magma, explosive and non-explosive respectively, and leading to the
formation of tuff deposits, may produce different types and amounts of
contamination in the deposits thus formed. In the explosive type there
is usually more than one outburst, but the tuff deposits formed from
outbursts spaced from each other by only a few minutes or hours are
sometimes difficult to separate from one another. The first phase is
often the most violent and can be recognized by the shape of the vesicles
in the small pumice fragments produced. These vesicles are aligned
parallel to each other and are greatly elongated in one direction, the
result of vesiculation within the conduit where the magma is highly
constrained and the gas vesicles can expand in only one direction, the
direction of motion. Such vesicles may be only a few microns in diameter
but a centimeter or more in length. In small sand-sized grains of pumice
they appear under a handlens or binocular microscope as fine striations
when the grains are viewed perpendicular to the direction of elongation
of the vesicles.

During this first eruptive phase, the orifice of the volcanic conduit
is rapidly enlarged by abrasion, and this abraded rock is added to the
rising cloud of debris. At Novarupta in Katmai National Monument,
Alaska, the conduit of the 1912 eruption passes through several thousand
feet of dark shale of the Naknek formation. Fragments of this dark
shale are easily distinguished in the white rhyolite pumice and ash of
the first phase of the 1912 eruption and can be found at distances of
20 to 30 miles from the orifice. Significantly, however, at distances of
100 miles or more from the orifice where the ash is, of course, much finer
grained, there is virtually no shale contamination owing to the fact that
most of the abraded shale fragments are large and were winnowed out
before travelling this distance. Granite lies below the Naknek formation,
and a few blocks up to several feet across may be found in the deposits

near the orifice, but granite fragments are difficult to recognize in the deposits a few miles away because of the similar color of the granite and rhyolite. This indicates the danger of contamination of this type going unrecognized in fine-grained deposits. In the explosive first phase of the Katmai eruption all of the contamination occurs as completely separate fragments of shale and, more rarely, granite. That is, no glass adheres to these foreign fragments and they are easily separable from the pumice and crystals of the finer parts of the ash using a binocular microscope. The possibility that some of the granite adjacent to a conduit might be broken into its component minerals must always be considered. Distinguishing fragmental potassium-feldspar of granitic origin from potassium-feldspar of volcanic origin may often be almost impossible; however, the presence of granitic feldspar in a sample may sometimes be established beyond reasonable doubt even when it cannot be recognized, if other more easily identifiable component minerals of the granite can be found. At Katmai, for instance, the rhyolite pumice contains only quartz and feldspar crystals but no ferromagnesian minerals such as biotite or hornblende, both of which occur in the granite. The absence of either biotite or hornblende in crystal concentrates of this ash indicates beyond reasonable doubt that the ash is uncontaminated by potassium-feldspar also.

It is not always necessary to discard a tuff sample in which contaminating older grains are recognized, because often the maximum age obtained from the contaminated sample may still serve a useful purpose in placing an upper limit on the date of the event. Sometimes, too, it is worthwhile to determine the age of the contamination itself, for if it is not much older than the sample, a few grains of it will not affect the age of the sample appreciably. For example, to see what effect a few undetected grains of contaminating minerals from the granite at Katmai would have on a K/Ar date of the ash, a sample of the granite taken from a boulder in the ash was dated. The granite has a geological age older than Upper Jurassic, as it lies unconformably below the Naknek formation of Upper Jurassic age, which would make the granite at least 150 million years old. The age obtained, however, is 4 million years! A similar age was obtained from a granite boulder in the Bishop Tuff in California.

Whereas small amounts of such contamination may have trivial effects on a K/Ar date, large amounts can play havoc, as was the case in the dating of the Bishop Tuff (EVERNDEN et al. 1957). We first determined the age of the Bishop Tuff to be approximately 950,000 years using a concentrate of potassium feldspar we suspected might be somewhat contaminated; however, we thought the temperature of the tuff itself was so high at the time of consolidation that all grains of contamination would be thoroughly degased. The later dating of the granite boulder suggested that this assumption might not be so, and DALRYMPLE (1965) subsequently prepared sanidine concentrates from pumice fragments associated with the tuff and obtained several dates all close to 700,000 years, a figure which is important for its control on the age of one of the early Sierran glaciations and for its control on the time of the second reversal, going back in time, of the earth's magnetic field which occurred later than the tuff but earlier than approximately 1,000,000 years.

At Katmai, at Bishop, and at many other volcanic centers, the first violently explosive phase of eruption was followed closely by outpouring of tremendous quantities (1 to 100 or more cubic miles) of particulate lava in a quasi-liquid state, the individual particles of which seem to be separated from each other and suspended in a compressed gas. These masses move out from their conduits and fissures adjacent to their conduits at high speeds as compared to the velocities of ordinary lava flows. While principally directed by local topography and gravity, they seem in part to be mobilized by the expanding gases within them and to have the ability to override obstacles in their paths of considerable height, nevertheless, the bulk of these eruptions stay close to the ground. A large cloud of ash accompanied this stage of eruption at Katmai and carried fine ash far beyond the limits of the main bulk, and doubtless such clouds accompany all eruptions of this type. When these huge masses finally come to rest, they often have sufficient residual heat to allow agglutination of their glass particles except along the margins and tops of the flows, and they form a welded mass having the superficial appearance of a lava flow.

It is clear that contamination in tuffs of this kind in both the welded and unwelded parts can be of two types: contamination from the walls of the conduits and contamination from debris incorporated in the ash as it moves violently over the earth's surface. Only the central part of the tuff-flow at Katmai is welded: the upper one hundred feet, the margins, and the last mile or two of the terminous of the tuff are unwelded and contain numerous fragments of igneous and sedimentary rocks of a variety of types, some of which being water-rounded, show clear evidence of having been picked up from river beds. Temperatures of fumaroles eminating from the tuff-flow were as high as 645° C seven years after the eruption in 1912 (ALLEN and ZIES, 1923) leading one to believe that contaminating fragments in old welded tuffs should be well degased; however, the experience of the Bishop Tuff proves how cautious one must be in such assumptions! Locally, large quantities of granite contaminate the Bishop Tuff (GILBERT, 1938) most of which probably represents conduit material. Many welded tuffs, however, exhibit very little contamination of this type and yield dates in concordance with associated air-borne tuffs. Of the two general types of contamination possible in welded tuffs, conduit wall-rocks and surface erosional debris, the first, if granitic in composition, is the most difficult to detect because of the general similarity in mineralogical composition and color of granite to most welded-tuffs themselves. As DALRYMPLE (1964) points out, whenever possible, mineral concentrates for K/Ar dating of welded-tuffs should be made from pumice fragments contained in the tuff to minimize the possibility of contamination.

Of the various erupted igneous material used for dating, lava flows, owing to their usually more quiet form of emission from vents, are probably least subject to severe contamination by incorporation of foreign material, although fragments of foreign rocks can frequently be seen in them and occasionally are very abundant. As shown by DALRYMPLE (1964), fragments of 100 million-year-old granite embedded in a basalt flow retained significant quantities of radiogenic argon which would have strongly affected the date of the basalt had the granite been over-looked.

Returning now to contamination in tuffs or tuffaceous-appearing deposits formed by frictional processes as described earlier, we can immediately see that the likelihood of contamination in them is very high from both the abraded conduit walls and from erosional debris in the stream channels over which they pass. When the process of breccia-tion that produces them occurs within the conduit itself the adjacent wall-rocks are strongly abraded, and the abraded and pulverized mater-ial is thoroughly mixed with the rest of the breccia in a short time (CURTIS 1954). While one can easily pick out fragments of the predominant lithologic type from the breccias and safely assume they represent primary material from the time of the eruption, the fine-grained matrix material is a homogenious mixture of primary and contaminating crys-talline and glassy debris. It is this fine-grained matrix material, separated from the coarser fragments by erosional processes and redeposited else-where, that closely simulates air-borne ash formed by explosive volcanism.

Contamination from reworking of primary tuffs by erosional processes

To be preserved in the geologic record, a volcanic tuff or lava flow must be deposited under conditions where it will be buried by other material before it is removed by erosion. These conditions are usually but not always best met in aqueous basins of deposition. In volcanic regions on land, the constructional process of volcanism may dominate the destructional processes of erosion, and volcanic ashes may be depos-ited and quickly buried by subsequent volcanic products before the soft ashes are stripped off by erosion. Because of the unconsolidated and

fragmental nature of air-borne ash and because it is deposited indiscrim-
inately on all land surfaces adjacent to the volcanic source, it is usually
quickly removed by erosion from most land surfaces and carried by
streams to the nearest aqueous basin of deposition. If such a basin be near
the volcanic source, it will usually already have received a layer of the
air-borne ash, and this layer will then be immediately buried beneath
a deposit of the reworked ash. Obviously, this reworked ash is more
likely to be contaminated by erosional debris than is the primary ash
layer at its base. Often the reworked part of a deposit of tuff is many
times thicker than the primary part. Sometimes the wind direction at
the time of eruption of an ash is opposite to the stream-drainage direction, in which case the local basin of deposition may receive only reworked ash; thus it is not always safe to assume that the basal layer of
a tuff is primary. In the first cataclysmic eruption from Novarupta at
Katmai, the wind direction was from the north through the first few
thousand feet of air, but above 30 or 40,000 feet it was apparently from
the south, for fine ash from this first phase reached over a hundred miles
to the north. Long before this fine ash settled, however, north-flowing
ash-choked streams from Katmai and Novarupta had carried huge
quantities of ash to Naknek Lake. It is safe to guess that cores of the
deposits in Naknek Lake, only 20 miles north of the source of the huge
1912 eruption, would show no primary uncontaminated ash layers from
the numerous phases of that eruption.

The criteria for recognizing primary tuff deposits must be used with
caution because they are sometimes applicable to reworked tuff deposited
close to the volcanic source. Similarly, the criteria for recognizing reworked tuffs are often applicable to primary tuffs deposited at long
distances from their sources.

Primary volcanic ashes deposited on land or in water close to their
sources are often massive without traces of bedding, and they usually
exhibit poor sorting as to size. If deposited in water, the pumiceous
fragments in an ash may settle more slowly than the mineral grains or
fragments of glass, hence these grains may be more concentrated in the
lower part of the ash-layer. The inclusion of abundant material of non-
volcanic origin—leaves, wood-fragments, bones, rounded pebbles or grains
of sand — in a tuff deposited in an aqueous environment is the most obvious criterion of its reworked nature. Reworked ashes nearly always show
marked stratification lines in their upper parts,and they are often cross-
bedded. Owing to multiple phases of an eruption, primary ash deposits
can also show stratification lines, but usually each stratum contains the
more or less unsorted components of that eruptive phase. Often in fine-
grained reworked tuffs, rounded grains can be detected only after the
desired mineral has been concentrated. Experience is necessary to distin-

guish rounding by stream transport from rounding by resorption of the grains while still in the magma: resorption often produces embayments in the grains while abrasion almost never does. If one suspects a reworked deposit of ash owing to the presence of a few rounded grains of feldspar in the concentrate, it is worthwhile to examine also the fraction of dark minerals. Here, the presence of metamorphic minerals or minerals known to be absent from the eruption may settle the question. Sometimes an obviously reworked tuff can be safely used if it contains a distinctive type of pumice from which a mineral concentrate can be made.

Summary

Contamination of samples is the single most important obstacle to obtaining both precise and accurate K/Ar dates from young geologic rocks. The answers to a variety of geologic problems can only be obtained by utilizing the ultimate resolving power of the K/Ar and other radiometric methods, and the presence of mere traces of contamination in the samples used may negate the results.

Contamination can occur geologically in a variety of ways, and the amount of contamination in a particular sample can depend in part on the mode of origin of the material used. Lava flows are usually the least contaminated and derived or reworked tuffs the most contaminated. Tuffs formed from air-borne ashes are usually less likely to be contaminated than welded tuffs, although mineral concentrates made from pumice fragments in either of these types of tuffs are generally the best of all materials for dating. All samples, of whatever origin, must be examined for contamination both in the field and in the laboratory during every stage of preparation.

Tektites

By

O. A. SCHAEFFER

I. Introduction

Tektites are small pieces of glass ranging in size from a few millimeters to the order of ten centimeters. The average size tektite is perhaps 2—3 centimeters in diameter. In general tektites show the appearance of having at one time been molten. Hence their name from the Greek word for melted. The molten character may be shown by tear drop shape and rounded rather than angular surface features. The first tektites to be described in the literature are those found in Czechoslovakia. The

Czechoslovakian tektites, generally called moldavites, were described by MAYER (1788).

For perhaps 100 years after MAYER's description of moldavite glass, it was considered to be a type of obsedian. In 1898 SUESS proposed that tektites, as he named the moldavite glass, were of extraterrestrial origin. From that date to the present time the argument has raged as to the origin of the tektite strewn fields.

In a superficial way the shape of tektites is not too different than one might expect for molten glass ejected by a rather violent volcanic explosion. However, on closer examination of the surface features of tektites, especially when studied carefully, there are clear indications that the surface features were formed during hyper-velocity travel through an atmosphere.

Chemically tektites are a high silica glass. The chemical composition of tektites is typical of the differentiated crust of the earth, that is, they have a relatively high alkali content with relatively low iron and magnesium. The aluminum content is also high. This similarity in chemical composition to a differentiated planetary surface is one of the strongest arguments against the origin of tektites either on the moon whose surface is not thought to be differentiated or from meteorites which are known not to be differentiated. If, on the other hand, the moon had been molten through much of its history, the surface rocks of the moon and many of the surface features of the moon are then probably the result of volcanic and tectonic activity. In this case, the surface of the moon may also be differentiated material similar to the crust of the earth. That is differentiated in the sense that through a long period of time, billions of years, the process of fractional crystallization has resulted in the concentration near the surface of minerals high in aluminum and the alkalies. The low density minerals differentiated from the high density minerals in the gravity field of the moon.

One of the main features of tekties is that they occur in strewn fields. Tektites occur in four rather large strewn fields at several locations on the earth's surface, as shown in Table 1. The major strewn fields occur in (1) Czechoslovakia, (2) southern United States, Texas and Georgia, (3) Australia, and southeast Asia and the Philippines and (4) the Ivory Coast. These strewen fields are rather extensive in origin. The strewn field in Czechoslovakia, for example, is perhaps two hundred kilometers in diameter. The one in Australia covers practically the entire continent. Yet other portions of the earth's surface are rather completely free of such objects. For example, no known tektites have ever been found in South America. The strewn field nature of tektites has led many to hypothesize that tektites are connected with some relatively local cataclismic event. Whether that event was the impact with the earth's surface by a large meteorite sloughing off material as it grazed the surface of the atmosphere, or the splashes from a major meteorite impact on the moon, ejecta reaching the earth in the form of tektites and being focused by the gravitational field of the earth—moon system or, the ejecta thrown out by some extremely violent volcanic event on the surface of the earth. All proposed origins of tektites share in common one particular feature, that is, a cataclysmic, event. It is probably for this reason that so much interest and attention has been focused on tektites. It is almost universally accepted that because of the extensive nature of the strewn fields, yet rather localized in global proportions, that some cataclysmic events have taken place on the surface of the earth in past times to which tektites may well furnish the only clues.

We can safely say that tektites were undoubtedly formed by some rather large melting of siliceous material. This corresponds to a precise event in time and one can attempt to isotopically measure the date of such events. If one wishes to date the event one needs an isotopic method where a daughter isotope is completely separated during this melting procedure. In principle, the best method which will be one which relies on the daughter being a noble gas which will very likely be completely de-gassed during the rather severe melting at the time of origin of the tektite material. Two such methods exist, one based on helium accumulation from the decay of the uranium and thorium chains and the other based on the accumulation of argon from the decay of potassium. Because of the high silica content of tektites it is unlikely that tektites will retain helium and furnish a method of measurement. On the other hand, the accumulation of argon furnishes a satisfactory method of age determination.

The potassium content of tektites is relatively high being of the order of 2%. With modern methods of mass spectroscopy, described in an earlier chapter, it is possible to measure the argon trapped in a tektite and obtain a reliable age for a sample as young as several hundred thousand years.

II. Experimental technique for application of K-Ar method to tektites

With modern mass spectrometric methods it is possible to determine quantities of argon as low as 10^{-9} to 10^{-10} cm³ STP. This means that samples which contain a few percent potassium can be dated even though the age may be only tens of thousands of years. The only limitation at present appears to be the amount of atmospheric argon associated with the sample itself. In determining the radiogenic argon content of tektites, it has been found to be of great advantage to use rather large pieces of tektite in order to keep the amount of atmospheric argon at a minimum.

As tektite glass has been very severely melted at some time during its history, it differs rather remarkably from obsedian or volcanic glass in its extremely low water content. As a result when tektite glass is melted, in order to release radiogenic argon, the glass in general melts without frothing or bubbling. For this reason, tektite glass is much more difficult to be outgassed than an obsedian. In general it is necessary, in order to be sure of total release of radiogenic argon, that the tektite be heated sufficiently so that it is distilled from the crucible. In this way the radio-genic argon is released during the volatilization process. If this procedure is not followed, one obtains erroneously low potassium argon ages.

Potassium argon ages of tektites have been determined extensively in two laboratories. One of these in Berkeley by Professor JOHN REYNOLDS and his students and the other one in Heidelberg by Professor WOLFGANG GENTNER and his co-workers. While there have been discrepancies for some of the early measurements, at present the results from the two laboratories are in excellent agreement. In the following paragraph these results will be considered for each tektite strewn field.

III. Potassium argon dates of tektite strewn fields

1. Austro-Asian Strewn Field. The largest strewn field of tektites is that which occurs in Southeast Asia, the Phillippines, and Australia. In this strewn field tektites are found in great abundance. Tons of material have been accumulated and exist in museums and private collectors possesions. This strewn field is the largest of all the strewn fields in exist-ence and extends for over 8 thousand kilometers. As will be seen in what follows, it is also the youngest strewn field in existence. As a result, one cannot be absolutely certain as to whether this strewn field is really larger than the other strewn fields or whether because of its relatively recent occurence it is more in evidence than the other strewn fields. The Austro-Asian tektites also show surface features which are not present on the other tektites. These morphological features consist of evidence of sculpturing of the tektite during flight through the atmo-sphere. The Australian tektites are the only tektites which show the delicate flanges and button shapes. This is often pointed out as a distin-guishing characteristic of australites. While there is little doubt that a tektite with a flange is almost beyond question an Australite, there is a good deal of question as to whether other tektites might not have been at one time similarly shaped. The flange material in general shows evidence of re-melting so that if such a tektite were buried it is quite likely that the flange material would decompose and be lost in times of the order of millions of years. The possibility exists then that the other tektite strewn fields may have at one time contained similar tektites

which during the course of time have lost these morphological features. The potassium argon results for tektites from this area are summarized in Table 1. It can be seen that the ages range from about 0.68 to 0.75 million years with an average of 0.7 million years. The ages determined are independent of locality whether they come from Australia, Indo-China, the Dutch-West Indies or the Philippines. It would thus appear, as has been suggested by ZÄHRINGER (1963), that the Austro-Asian tektite field is a single field of relatively young age, geologically approximately mid-pleistocene. As the error of the measurements are of the order 100,000 years, it is of course not possible to distinguish between a single event which produced all these tektites, or a number of individual events spaced over a period of several hundred thousand years. It would appear, however, from the potassium argon results that those events, if multiple, must have taken place within approximately one hundred thousand years of each other.

According to BAKER (1963) australites occur close to the surface of the earth in loose recent sediments. According to FENNER (1935) no australites have been found in any formation except aeolian or alluvial drifts and as a result FENNER concludes that they fell in early post pleistocene times. Australites are also found on mountain ranges, plateaus, hilltops and surfaces of volcanic plains. They come from the indigenous surface and soil of the earth from sea level to heights over 2,000 feet. As a result, BAKER concludes that the Australian tektites were emplaced approximately 5,000 years ago, probably not under 3,000 years nor over 6,000 years.

In the case of the australites there is apparently a direct conflict between the geological age and the potassium argon age of the specimen. Perhaps the greatest potential source of error in K-Ar dating lies in the fact that the tektites might contain inherited argon. That is, at the time of their formation when they were molten not all the argon was removed. As a result, tektites would be expected to show somewhat older ages than their true age. On the other hand, as tektites in general are detrital minerals in sedimentary deposits it is not unusual that they should be older than the deposit itself. It would be unusual for them to be younger than the deposit.

As the potassium argon age measures the formation age of the tektite, that is the time that has elapsed since the tektite was a molten droplet and the geological age measures on the other hand the time on earth another possible explanation for the apparent difference in ages is that tektites originate from an extra-terrestrial source and that the difference in the two ages is accounted for by the time required to reach the earth from their extra-terrestrial place of formation. There is some evidence which would indicate that this is not the reason for the apparent difference in the geological age and the potassium argon age. If tektites originated on the surface of the moon or had some other extraterrestrial meteoritic origin it is difficult to see how they could have come from the moon or some other place in the solar system and not been exposed to cosmic rays as is well know to be the case for meteorites. If this roughly half million year difference in age were spent in space, evidence of cosmic ray irradiation should be found. There is clear evidence that such is not the case. REYNOLDS (1960e) has shown that australites do not contain detectable neon-21. From the measured neon diffusion rate, the measured cross section for the production of neon from light elements, and from the limit of detectability of neon-21 REYNOLDS concludes that the australite he measured had a maximum flight time of 28,000 years. ANDERS (1960) does not find any aluminium-26 in an australite. The sensitivity of the measurement of aluminium-26 (VISTE and ANDERS, 1962) does not, however, preclude a 90,000 year flight time.

There thus remains three possible explanations for the discrepancy between the geological and the potassium argon age of the australites: (1) The geological age does not measure the arrival of the tektites on the earth as the tektites are detrital to the deposits and were originally deposited in other older formations. (2) The tektites contain inherited argon which make the measured potassium argon age too old. (3) The tektites spend approximately 0.5 million years in transit between their places of extra-terrestrial origin and their arrival on the earth. In this case, the lack of neon-21 in the tektites would indicate that either the tektites had lost their neon due to severe heating during their flight through the atmosphere or that the tektites travelled in space as a large swarm and in this way were shielded from cosmic radiation.

According to LACROIX (1935b) and BAKER (1963) loc. sit. the Muong-Nong tektites from Indo-China occur over a distance of 800 kilometers in northwest-southeast direction and over a distance of 500 kilometers in a southwest-northeast direction. The presently known distribution of Indo-Chinites covers distances of over 1600 kilometers from Surat Thani in Thailand to the island of Tanhai off the coast of China and as far south as Dalat in South Viet Nam. Several specimens have been found at a depth of over 1000 meters in the South China Seas, 270 kilometers off the Viet Nam coast. SAURIN (1935) states that tektites in Indo-China rest on all formations exept recent alluvial. As a result Barnes concludes that the time of fall of these tektites should not be later than 10 to 15 meter terrace of quaternary age in southern Indo-China. In Kwang Chowan, Indo-China the tektites fell to the earth after the formation of the ancient dunes and before the outpouring of the gravel.

It would thus appear that the Moung Nong tektites geologically are older than the Australian tektites. On the basis of potassium argon dating such a difference in age is undetectable and if it exists it should be less than 100,000 years. From geological evidence as described by BARNES, it would appear that the tektites from Dilitin Island are of a similar age to the Indo-Chinites.

The tektites from the Philippines have been described by BEYER (1955). According to BARNES, loc. cit., the age of Philippinites should be somewhat intermediate between those of the indo-chinites and the australites. However, the geological evidence in the case of the Philippinites leaves much to be desired. Most of the tektites collected by BEYER's were from quarry waste on a quarry floor and their association with the overlying laterite is surmised. An age of 0.5 million years would not be discrepant with the geological evidence in Indo-China, Malaysia, or the Philippines. There is no clear evidence that perhaps the tektites in this area are a series of events rather than a single event.

What can be said without doubt is that the tektites found in Australia, Philippines and Southwest Asia were all formed of the order of half a million years or less ago. This is incontravertable on either geological or potassium argon dating grounds. Whether all these tektites were deposited simultaneously that is within a year or less of one another, or whether they were deposited over a span in time by several discreet events, the potassium argon ages would say that if such were the case then these events took place over a period of less than 200,000 years at a time 700,000 years ago.

2. The Ivory Coast Tektites-Potassium argon ages indicate an age, see Table 1, of 1.2 million years for the Ivory Coast tektites. The geological setting, according to BARNES, would only indicate that Ivory Coast tektites are relatively young. They occur in gold bearing alluvial which rests on very ancient crystalline rocks. It is extremely improbable that they have been weathering from the ancient crystalline rocks. For this reason they must have been in the zone of weathering since they fell.

3. Moldavites: The moldavites were the first tektites observed. They give a potassium argon age as indicated in Table 1 of 14.9 ± 0.1 million years. JANOSCHEK (1934) concludes that the moldavites in Moravia are associated with a gravel of Helvetian age. The potassium argon age of slightly less than 15 million years is not in disagreement with such a stratigraphic assignment.

4. Bediasites. — The bediasites are the tektites found in Texas. They show an average age as shown in Table 1 of 34 million years. The Texas tektites are associated with the Jackson formation which is upper Eocene. The Eocene lies between 35 and 55 million years ago so that the potassium argon age is in good agreement with the stratigraphic asignment. The Georgia tektites show an age indistinguishable from the Texas tektites. They reportedly come from the Hawthorne formation of middle

Table 1. *K-Ar ages of tektites*

Sample Description	%K	Ar[40] rad. 10^-8 STEc m³₁g	K-Ar age m. y.	Ref.
Australites:				
Kalgoorlie, W. Australia .	1.95	3.4	0.44 ± 0,2	[1]
Kalgoorlie, W. Australia .	2.00	6.3	0.80 ± 0.2	[1]
Australia . . ·	1.66	4.0	0.61	[2]
Charlotte waters,C.Australia	1.88	5.1	0.69 ± 0.06	[5]
Wiluna, W. Australia . . .	1.99	5.9	0.75 ± 0.06	[5]
Oodnadatta, S. Australia .	1.76	5.3	0.73 ± 0.05	[5]
Florieton, S. Australia . .	1.94	5.6	0.73 ± 0.05	[5]
Pine Creek, S. Australia . .	1.90	5.7	0.76 ± 0.06	[5]
Australit	1.90	5.1	0.68 ± 0.06	[5]
Indochinites:				
Tan Hai, China	2.13	6.6	0.79 ± 0.2	[1]
Sumatra	1.76	3.5	0.5 ± 0.2	[1]
Tan Hai, China	2.04	4.1	0.51	[2]
Kwangchowan	2.06	5.2	0.65	[2]
Kwangchowan	1.89	4.7	0.63	[2]
Dalat Annam	2.04	4.7	0.59	[2]
Dalat Annam	2.25	5.4	0.62	[2]
Kouang Tscheon Wan . .	1.68	4.8	0.72 ± 0.06	[5]
Dalat Annam	2.03	5.9	0.73 ± 0.07	[5]
Kambodga	2.09	6.0	0.73 ± 0.07	[5]
Thailand	1.85	5.1	0.71 ± 0.06	[5]
Sangiran, Java.	1.63	4.7	0.73 ± 0.05	[5]
Bruneit, Borneo	1.82	5.3	0.73 ± 0.05	[5]
Bilitonites:				
Billiton Island	2.11	4.7	0.57	[2]
Billiton Island	2.10	6.0	0.72 ± 0.08	[5]
Philippinites:				
Santa Mesa	2.29	5.5	0.61	[2]
Santa Mesa	1.95	5.4	0.69	[2]
Bulacan.	1.84	4.4	0.62	[2]
Pugad Babuy	1.80	4.7	0.66	[2]
Santa Mesa	2.10	5.7	0.69 ± 0.06	[5]
Santa Mesa	2.29	6.6	0.73 ± 0.07	[5]
Pugad Babuy	2.00	5.4	0.69 ± 0.07	[5]
Anda	2.15	5.8	0.68 ± 0.11	[5]
Santiago Isabella	2.01	5.8	0.73 ± 0.06	[5]
Moldavites:				
Locenice	2.72	160	14.9	[3]
Radiomilice	1.96	114	14.6	[3]
Habri.	2.67	154	14.5	[3]
Moravia.	2.12	123	14.6	[3]
Lhenice	2.60	153	14.9 ± 0.7	[5]
Dukovany I	2.49	136	13.8	[3]
Dukovany II.	2.68	160	15.0	[3]
Moravia.	2.49	120	14.7	[3]

Table 1. Continued

Sample Description	%K	Ar⁴⁰ rad. 10⁻⁸S TP cm³/g	K-Ar age m. y.	Ref.
North America:				
Lee County, Texas	1.46	204	35.0 ± 2.0	[5]
Lee County, Texas	2.12	248	29.0 ± 0.9	[1]
Grimes County, Texas . .	1.60	215	33.8 ± 1.9	[5]
Grimes County, Texas . .	1.65	221	33.7 ± 1.9	[5]
Gonzales County, Texas. .	1.68	230	34.4 ± 2.0	[5]
Fayette County, Texas . .	1.71	233	34.2 ± 2.2	[5]
Fayette County, Texas . .	1.59	216	33.4 ± 2.6	[5]
Georgia	1.99	267	33.9 ± 1.7	[5]
Martha's Vineyard	2.00	268	33.7 ± 2.2	[5]
Ivory Coast Tektites:				
Ouellé	1.33	6.5	1.2 ± 0.2	[5]

[1] REYNOLDS (1960e), [2] GENTNER and ZÄHRINGER (1960), [3] GENTNER, LIPPOLT and SCHAEFFER (1963), [5] ZÄHRINGER (1963).

or lower Miocene age (COOKE, 1943). The Miocene lies roughly between 11 and 25 million years ago, so that lower Miocene age would not correspond to potassium argon age found. Recently (CLARKE and HENDERSON, 1961) have pointed out that the conclusion that the Georgia tektites come from the Hawthorne formation is somewhat doubtful. As a result, there is not necessarily a discrepancy between the K-Ar age and the stratigraphic evidence. The one tektite from MARTHA's Vineyard gives an age similar to the Bediasites. As it is a single example there is some doubt wether its occurence marks a tektite field centered on New England.

IV. Potassium Argon ages, and the origin of tektites

It has been pointed out by numerous authors that the occurrence of tektites in strewn fields is indicative of there being several discreet events causing strewn fields of tektites. Occasionally there have been suppositions that tektite strewn fields, for example, that in Texas and that in Czechoslovakia were of simultaneous origin. The Potassium argon results are quite conclusive in this point. It would indicate that the Australian-Asian strewn field, the Czechoslovakian strewn field, the Ivory Coast strewn field and the North American strewn field were each discreet individual events occuring 34 million years ago, 15 million years ago 1.3 million years ago and less than 0.7 million years ago. The latter expressed as a limit to allow for possible contribution of inherited argon. The older ages should not suffer from any errors due to inherited argon because of their relatively large Ar⁴⁰ content.

The speculations on the origin of tektites go back to the earliest literature of roughly 170 years ago. The various hypothesis can be divided into two classes one of which favors a terrestrial origin for tektites and the other favors an extra-terrestrial one. The evidence has been summarized by O'KEEFE. The chemical composition of tektites is very similar to the chemical composition of the earth. The most obvious way to explain strewn fields (SPENCER, 1933) is that each represents the ejecta from a single event on the surface of the earth. The most convincing argument that has been raised against a terrestrial origin is an aerodynamic one. This argument is discussed in detail by CHAPMAN and LARSON (1963). According to them the surface sculpturing of the Australites indicates that the Australites entered the earth's atmosphere at a relatively low angle and at a relatively high velocity. From these considerations CHAPMAN concludes that Australites could not have originated on the earth's surface because aerodynamically it is impossible to throw a small object through the earth's atmosphere. The only way this can be accomplished is to have the meteorite impact and the subsequent shock waves reduce the atmospheric pressure to an extremely low value, something less than 10^{-3} times the normal value. CHAPMAN asserts that such is rather an unlikely occurrence. On the other hand, there have been few studies of the effects on the atmosphere of the impact of a comet or asteroidal size body with the earth.

V. Potassium Argon dating of glass from large meteorite impact craters

One method of shedding some light on the terrestrial or extra-terrestrial origin of tektites would be to locate possible crater sites which may be the meteorite or cometary impact site from which the tektites originated. If then the potassium argon date from all known craters were different from the potassium argon date of the tektite age, terrestrial origin would become extremely improbable. On the other hand if a number of craters gave potassium argon ages which agreed well with the potassium argon date of the tektites in the neighboring strewn fields then the probability of a terrestrial origin would be increased. It must be borne in mind that the potassium argon dates of two such samples can never prove the type of simultaneity that would be required to remove all doubt as to the origin of the tektites. Inspite of this, the dating of craters does furnish one additional piece of evidence in the fabric which will one day elucidate the origin of tektites. The terrestrial origin of tektites has been suggested from practically the earliest literature on the subject. A recent survey of the arguments in favor of terrestrial origin are given by O'KEEFE (1963). The hypothesis that the Czechoslovakian tektites originated as the result of a meteoritic impact with the earth was suggested by several authors. The largest known crater in the vincinity of the moldavites lies in Southern Germany, the Ries Crater, which is approximately 100 kilometers northwest of Munich. The small town of Nördlingen lies within the crater. The crater itself is about 20 kilometers in diameter and crater lies approximately 250 to 400 kilometers from the Czechoslovakian tektites. All of the Czechoslovakian tektite occurrences lie within a 15° angle from the Ries crater. Glass specimens from the outside rim of the Ries Crater have been dated by GENTNER and co-workers and give the results indicated in Table 2. The agreement between the seven different localities of glass is well within the error of the dating method and is in remarkable agreement with the average value for the Czechoslovakian tektites listed in Table 1.

Table 2. *K-Ar of crater glass*

Sample	%K	10^{-8} cm^3/g Ar^{40} rad.	K-Ar age m. y.	
Bosumtwi Crater - Ghana [1]:				
Buonium river	2.3	9.6	1.2 ± 0.1	
Ata River	1.53	8.6	1.4 ± 0.2	
Nördlingen Ries Crater - Bavaria [2]:				
Otting I	3.10	186	15.1	
Otting II	2.91	173	15.0	
Zipplingen	2.18	130	15.2	
Ottingen	2.50	144	14.5	
Bollstadt	2.32	125	13.6	
Altenburg	2.69	156	14.6	
Aufhausen	2.16	130	15.2	
Mauren	2.42	149	15.5	

[1] GENTNER, LIPPOLT and MÜLLER (1964), [2] GENTNER, LIPPOLT and SCHAEFFER (1963).

In a similar way the crater at Lake Bosumtwi in Ghana is related to the Ivory Coast tektites. In this case the Ivory Coast tektites are approximately 300 kilometers from the Crater and all lie within a 12° angle. The crater itself is smaller than the Ries Crater; the Bosumtwi Crater is about 10 kilometers in diameter. Glass from this crater has been dated by GENTNER et al. (1963) and compared with the postassium-argon dates of Ivory Coast tektites. In this case, again, the agreement between the crater glass and the tektites is excellent.

There is no good candidate for a crater for the Australian tektites or the North American tektites. On various occasions Antarctica has been suggested as a site for the Australian tektites and some locality in Canada such as Hudson Bay as a possible site for the Texas tektites. This would place both these tektite fields extremely far from their place of origin, much further than is the case for either the Ivory Coast or the Czechoslovakian tektites.

Were it not for the aerodynamic argument of CHAPMAN one would be tempted to say that this agreement between two craters and the tektites from two strewn fields is evidence enough for the terrestrial origin of tektites. Especially in view of the fact that the chemical composition of tektites speaks strongly for origin by the melting of a differentiated crust. However, according to CHAPMAN the atmosphere precludes the association of a tektite strewn field and a crater of any great distance. In addition, the sculptured features of the Australites would indicate that they came into the earth's atmosphere from outside and were sculptured by hyper-velocity flight during the return transit through the atmosphere. This argument also speaks against the formation of Tektites from any planetary surface, with even a tenuous atmosphere. If the surface of the moon is not differentiated, then on purely chemical grounds the moon will also be eliminated as a possible source of tektites.

In summary then potassium argon dates have proved of significance in the understanding of tektite strewn fields. They have shown conclusively that each strewn field is a different entity and that the tektite strewn field in Czechoslovakia is not related to that in Texas as previously suggested, so that the arrival of tektites at the earth is a more or less local event rather than a global event.

In addition, K-Ar dates have shown that there are two remarkable coincidences between the dates of large meteorite impact craters and tektites in neighboring strewn fields. These coincidences strongly suggest that these tektites are for these cases formed by meteorite or cometary impact with the surface of the earth and that in some way either during the meteorites arrival, the shock wave afterwards, or by the ejection of a large mass of material, the tektites were able to be thrown out of the earth's atmosphere and then returned after being cooled sufficiently to solidify. Undoubtly new evidence will be obtained when material is returned from the surface of the moon.

K-Ar Ages of Meteorites

By

D. KRANKOWSKY and J. ZÄHRINGER

A. Stone Meteorites

Introduction

The first attempt in dating meteoritic material was carried out by PANETH and his co-workers (ARROL et al., 1942). They tried to measure the U-He age of iron meteorites. An age determination was not possible, however, because the He contained a large amount of non-radiogenic He³ (PANETH et al., 1952). The previous suggestions by BAUER (1947) and HUNTLEY (1948), that helium is likely to be produced by the interaction of cosmic rays rather than by the decay of uranium, were verified.

Shortly after the discovery of the K-Ar method GERLING and PAV-LOVA (1951) measured the K-Ar ages of stony meteorites. After this first work the method was used by a number of authors (GERLING and RIK, 1955; WASSERBURG and HAYDEN, 1955a; THOMSON and MAYNE, 1955; GERLING and LEVSKII, 1956; REYNOLDS and LIPSON, 1957; GEISS and HESS, 1958) and it turned out, that stone meteorites have ages up to 4.5 b.y., which is much higher than the age of any terrestrial material. In the meantime a large number of data has been accumulated, which confirms the previous results and allows important conclusions for the history of meteoritic material and the development of the solar system to be drawn.

At the same time Pb-Pb ages (PATTERSON et al., 1953; PATTERSON, 1955) and Rb-Sr ages (HERZOG and PINSON, 1956; SCHUMACHER 1956a) of 4.5 b.y. were reported for stone meteorites. The measurement of the uranium concentration by HAMAGUCHI et al. (1957) made it possible to determine U-He ages of stone meteorites. In a recent work of HERR et al. (1961) the determination of Re-Os ages of iron meteorites has been attempted.

The first attempts of ZÄHRINGER and FIREMAN (1956) and STOENNER and ZÄHRINGER (1958) to date iron meteorites by the potassium argon method and later work by FISHER (1963, 1965), who used the same method, gave ages much higher than those of stone meteorites. Meanwhile these experiments have been continued by MÜLLER and ZÄHRINGER (1966) with a more refined technique. An age of 6.3 b.y. is the most likely K-Ar age for iron meteorites. In the last part of this chapter these new results will be presented.

The K-Ar ages as well as the U-He ages are gas retention ages and
as diffusion losses of argon and helium are strongly temperature dependent, they may reflect the thermal history of meteorites. The Pb-Pb
and Rb-Sr ages date the time of solidification of the material, when the
chemical differentiation ended. These ages reflect an earlier part of the
evolution of meteorites, when the meteoritic material was still incorporated in larger so called parent bodies and began to cool. Besides these
dating methods, which are based on the decay of long lived nuclei, there
are the radiation ages or cosmic-ray exposure ages, which date the break-
up and destruction of larger bodies in space by collisions and space
erosion.

When the K-Ar ages of a larger number of meteorites had been determined, it turned out, that together with the other methods, especially
with the exposure ages, one could arrive at more detailed conclusions.
It was recognized, that the age distribution of a large number of meteorites rather than the K-Ar age of a single object is informative (KIRSTEN
et al., 1963). This statistical interpretation of the data showed, that there
exists a characteristic difference between the age distribution of the
different meteorite classes. Differences also exist for the U-He and ra-
diation-age distributions. This article will demonstrate, that the K-Ar
method together with the other above mentioned methods is a powerful
tool to increase our knowledge of the history of the meteoritic material
and therefore of the planetary system.

1. The classification of meteorites

A short summary of the classification of meteorites according to their
chemical composition and their mineralogy will be given. The interested
reader is referred to the text-books of COHEN (1894, 1903, 1905), KRINOV
(1960) and MASON (1962a), to the monographies of TSCHERMAK (1885)
and of PERRY (1944) and to the review article of WOOD (1963). In a more
recent review article of ANDERS (1964) one finds also further references.

The most frequently used system of classification is that of ROSE
(1863), TSCHERMAK (1872), and BREZINA (1904) modified by PRIOR (1920).
In this scheme meteorites are divided into iron-meteorites (siderites),
stony irons (siderolites) and stone meteorites (aerolites) according to the
relative proportions of the metal and silicate phases. Stone meteorites
are further classified into chondrites (which contain small silicate and
other chondrules) and achondrites (without chondrules). According to
their chemical and mineralogical composition, the chondrites are sub-
divided into enstatite chondrites, bronzite chondrites and hypersthene
chondrites. The main criterion for this division is the iron content of the
silicate phase (mainly olivine and pyroxene), which increases from the

enstatite to the hypersthene chondrites. Two further classes of chondrites, the pigeonite and the carbonaceous chondrites, have been introduced. Pigeonites are also defined by their iron content in the silicate
phase (MASON, 1962b; MASON, 1963a). The carbonaceous chondrites
differ from the other ones by their relative high content of volatiles
(carbon $0.5-5\%$; $1-20\%$ H_2O e.g.) (MASON, 1963b). MASON and WIIK
(1964) have shown that the amphoteritic chondrites (low metal content
$1-5\%$; high FeO content in the olivine and orthopyroxene) form a
subclass of the hypersthene chondrites. The achondrites are divided into
calcium poor ($0-3\%$ CaO) and calcium rich ($5-25\%$ CaO) achondrites
(PRIOR 1920). A criterion for a further division is the relative proportion of FeO and MgO (MASON, 1962a).

Within the group of the iron meteorites one distinguishes on the
basis of their structur and nickel percentage nickel-poor ataxites, hexa-
hedrites, octahedrites, ataxites with a medium nickel content and nickel-
rich ataxites (PRIOR, 1920).

Iron meteorites are mainly composed of Ni-Fe alloys. They contain
small amounts of schreibersite (Fe, Ni, $CO)_3$ P and cohenite (Fe_3C). Graphite, troilite, sulfides, oxides and silicates are found in roundish to
oval inclusions (nodules) and veins embeded in the NiFe texture. Their
sizes reach from microscopical small particles to those of 10 cm in diameter or more.

EL GORESY (1965) distinguishes between 3 classes of nodules:
graphite-, graphite-troilite- and troilite nodules. Inclusions of silicates
have been found by MUELLER and OLSON (1964). As far as the K-Ar
dating of iron meteorites is concerned, one important finding is, that
graphite- and graphite-troilite nodules may contain up to 20% silicates.
The silicates are mainly composed of olivine (($Mg, Fe)_2 SiO_4$), clinoensta-
tite ($Mg_2Si_2O_6$), diopside (($Ca, Mg) Si_2O_6$), apatite ($Ca_5(Cl, F) (PO_4)_3$) and
feldspars (($Na, Ca) AlSi_3O_8$). Diopside, apatite and feldspar can contain
traces of potassium. This point will be discussed more extensively, when
the dating of irons with the K-Ar method is described.

2. Sources of Ar-isotope variations

For the calculation of the K-Ar age one has to know the K^{40} content
and the Ar^{40} content coming from the decay of natural occuring K^{40}.
Meteorites contain potassium as a minor constituent. The distribution
of potassium is not homogenous, because of the complex mineralogical
and chemical composition of the meteorites. Therefore the K^{40} and Ar^{40}
content has to be measured within the same sample of a meteorite specimen (cf. MÜLLER, this book).

There are four sources, which contribute to the total amount of Ar^{40} in the meteorites.

1. Radiocative decay of K^{40},
2. Cosmic ray-induced spallation reactions,
3. Primordial argon, which has been incorporated during the formation of the meteorites,
4. Contamination of terrestrial argon by surface adsorption.

In order to get the radiogenic Ar^{40} content, one has to correct for the argon from sources 2 to 4. The contribution from source 4 is easily to account for. As this argon is only adhered to the grain surfaces, it can be removed by heating the meteorite sample to 100 to 150° C before gas extraction (cf. KIRSTEN, this book). Degassing experiments have shown, that there are no essential losses of radiogenic Ar^{40} at these temperatures (ZÄHRINGER, 1962). Contributions of sources 2 and 3 are generally very small. Only for meteorites containing large amounts of primordial argon a small correction is needed.

Table 1. *Content and istopic composition of rare gases in some typical meteorites with schematic indication of their origin*

▨ *Radiogenic*
▩ *Spallation prod.*
☐ *Primordial*

	He^3	He^4	Ne^{20}	Ne^{21}	Ne^{22}	Ar^{36}	Ar^{38}	Ar^{40}
Iron meteorite *Sikhote–Alin*	.37	160	0.52	0.38	0.42	1.36	2.25	≼0.5
Ordinary chondrite *Mauerkirchen*	.59	915	8.7	9.2	10.4	1.31	1.39	5400
Enstatite chondrite *Abee*	12	1320	10	2.5	3.3	37.2	6.9	6900
Achondrite *Staroe Pesyanoe*	220	630000	2080	23	184	148	28	2700
Cosmic abundance *with* $Si=10^3$	?	3000000	7740	26	836	126	24	?
Terrestrial abundance *in ppm by volume*	$7·10^{-6}$	5.2	16.5	0.047	1.6	31.4	5.9	9300

In 10^{-8} *cc per g*

Primordial rare gases were first detected by GERLING and LEVSKII (1956) in the achondrite Staroe Pesjanoe. Later it turned out, that they are very common among stone meteorites. There are two sorts of primordial rare gases, an unfractionated component containing the rare gases in relative proportions and with isotopic composition similar to the cosmic abundances and a fractionated component, where the light rare gases He and Ne are depleted. The kind of rare gases found in a meteorite seems to depend on the chemical and mineralogical composition i.e. on its membership to a special meteorite class. More details and also references can be found in the review articles of SIGNER and SUESS (1963), ZÄHRINGER (1964) and ANDERS (1964). An important result of these studies is, that the unfractionated as well as the fractionated component contains argon with a Ar^{36}/Ar^{38} ratio equal within $\pm 5\%$ to the terrestrial ratio of 5.4. To correct for Ar^{40} from source 3 one has to know the relative primordial Ar^{40} abundance. As nearly all terrestrial Ar^{40} has been produced by the decay of K^{40}, the primordial abundance of Ar^{40} must be much smaller than in atmospheric argon. In meteorites Ar^{40}/Ar^{36} ratios as small as 5 have been found. A correction for primordial Ar^{40} is therefore usually not necessary.

The argon isotopes, which are produced as residual nuclei in the spallation reactions, derive from target nuclei with mass numbers larger than 40. In stone meteorites only iron, nickel and calcium are present in sufficient amounts. Since the cosmic ray exposure age of stony meteorites is so small, the amount of spallation produced Ar^{40} is also small compared to the amount of radiogenic Ar^{40}.

This situation is illustrated by Table 1, which shows the contributions of radiogenic, spallation and primordial rare gases to the total in some typical meteorites. The Ar^{40} concentration in stone meteorites ranges between 100 and 8000 $\times$ 10^{-8} cc STP per gram.

3. K-content and isotopic composition

Fig. 1 shows a histogram of the potassium content of stone meteorites. Only those data in the literature are used, which have been obtained by modern analytical techniques of isotope dilution, neutron activation

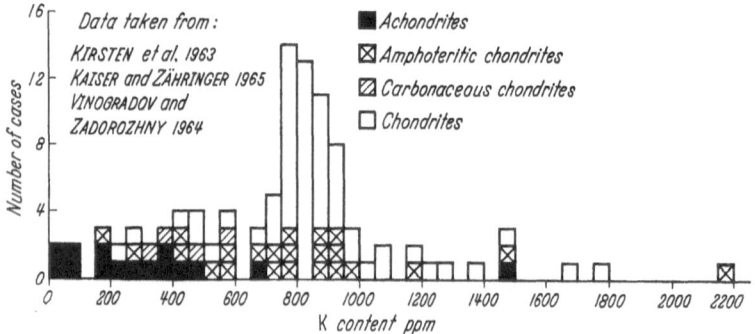

Fig. 1. Histogram of the K content of stone meteorites

and flamephotometry together with a sublimation technique for the extraction of potassium. The older data, where the classical methods of quantitative chemical analysis have been applied, turned out to be too high. This is due to contamination with terrestrial potassium (cf. MÜLLER, this book). Fig. 1 shows, that most of the chondrites (including enstatite-bronzite-, hypersthene- and pigeonite-chondrites) have potassium contents between 750 and 950 ppm. This interval becomes even narrower if one omits the results, which have been obtained by classical wet chemical extraction of potassium together with flamephotometry. Carbonaceous chondrites and achondrites tend to have lower potassium contents. The achondrites scatter widely from 10 ppm for Johnstown to 690 ppm for Stannern (omitting Shergotty with 1500 ppm). Amphoteritic chondrites show also a large scattering (200 ppm Ensisheim and 2180 ppm Soko-Banja). Contrary to the other chondrites they exhibit in addition large variations of the potassium concentration within a single meteorite. For example in Jelica variations from 560 ppm to 1170 ppm are found. Therefore the histogram includes in four cases a second potassium determination for the same meteorite. Possibly a potassium bearing mineral like feldspar, which tends to concentrate into "pools", is responsible for this inhomogenity (KAISER and ZÄHRINGER, 1965). For a K-Ar dating one can take an average value of 800−900 ppm for the enstatite-, bronzite, ordinary hypersthene- and pigeonite-chondrites. In all other cases the potassium and the Ar^{40} concentration must be measured in the same homogeneous meteorite sample.

For the calculation of the K-Ar ages, the assumption of identical isotopic composition of meteoritic and terrestrial potassium is made. The relative abundance of K^{40} is very small (0.0118%) and even 4.5 b.y. ago it amounted only to 0.13%. At this time some process might have taken place, which added potassium of different isotopic composition to the primordial potassium. This amount of additional potassium could have been different for meteoritic and terrestrial material. Such a process has been proposed by FOWLER, GREENSTEIN and HOYLE (1962). They argued, that the inner parts of the solar system have been exposed to an extensive particle irradiation during the time of their accumulation. This should result in different K^{40}/K^{41} ratios, if the relative amounts of irradiated material were different for the earth and the meteorites (BURNETT et al., 1965). Recently BURNETT et al. (1966) measured the K^{40}/K^{41} ratio in nine stone meteorites, one mesosiderite and one iron meteorite. They found the same value within $\pm 1\%$ for the earth and the meteorites, which excludes a modification of the K^{40}/K^{41} ratio by the Fowler, Greenstein, Hoyle mechanism. These results also seem to show, that no other process was effective in changing this isotope abundance. This conclusion can be compared with that of an older work of RIK and SHUKOLJUKOV (1954) and a very recent one of KEMPE (1965), who found agreement with the terrestrial K^{40}/K^{41} value for two other stone meteorites and also for inclusions in some iron meteorites. During the exposure to cosmic radiation K^{40} is produced in meteorites by spallation reactions. These amounts are neglegible compared to the primordial K^{40}. Only meteorites containing little primordial potassium (< 1 ppm) such as iron meteorites or potassium-poor mineral components have appreciable amounts of cosmogenic K^{40}. Burnett et al. (1966) measured K^{40} anomalies ranging from 1 to 1.5% in different mineral fractions of the Norton County

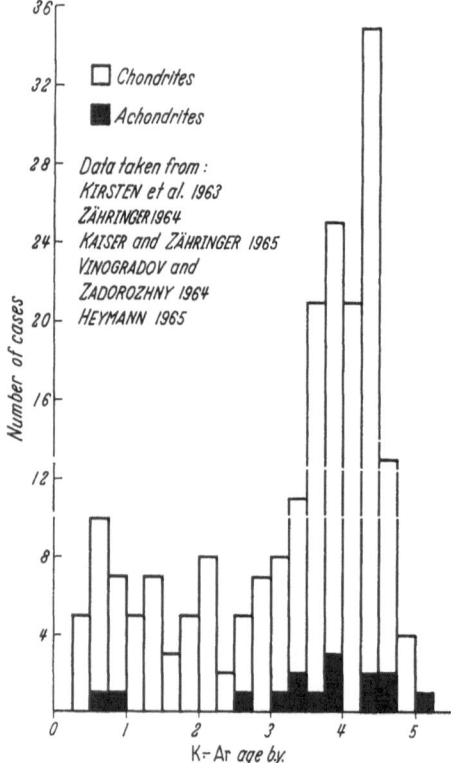

Fig. 2. Histogram of K-Ar ages of stone meteorites

achondrite, silicates from the mesosiderite and the iron meteorite. For Norton County the anomalies were correlated with the Ca content, which might be the target nucleus for cosmogenic K^{40}. This has been produced during the last 100 m.y. or less. Only 2.5% have been decayed up to this time, which would not have any significant influence on the K-Ar age. In summary one must conclude, that the isotopic composition of K in stone meteorites is the same as the terrestrial one.

4. K- Ar ages

After condensation and accretion into parent bodies the meteoritic material underwent chemical differentiation. The K-Ar method like the

other dating methods based on radioactive decay determines the time
at which the separation of mother- and daugther-element ended. An
advantage over the Rb-Sr, Re-Os and Pb-Pb-methods is, that the separation of mother- and daughter-element has been almost complete
which increases its sensitivity. The volatility of the daughter element,
however, implies also a disadvantage of this method. The K-Ar age is
much more accessible than the other methods to later changes by
diffusion losses of the daughter element.

Therefore the K-Ar age of a meteorite gives only a lower limit of the
time, when the last complete separation of potassium and argon took
place. On the other hand, these diffusion losses can give additional
information on the thermal history of the meteorites.

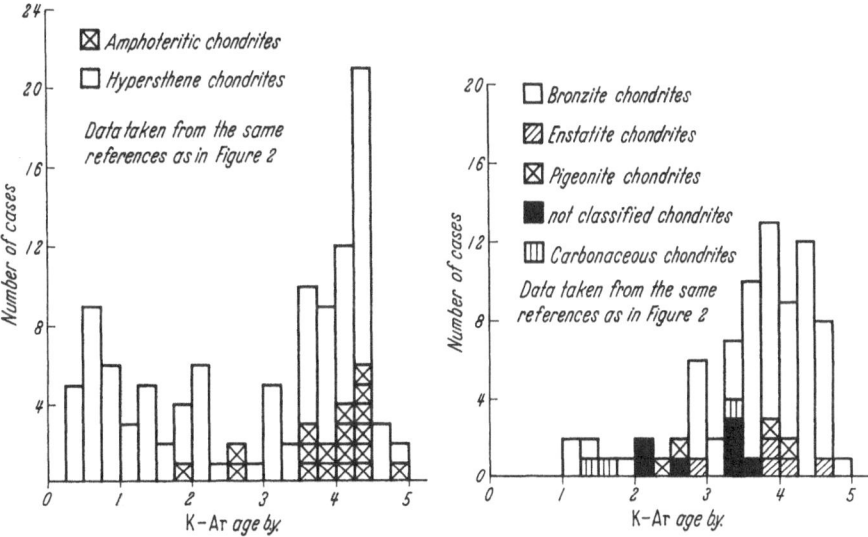

Fig. 3. Histogram of K-Ar ages of hypersthene
and amphoteritic chondrites

Fig. 4. Histogram of K-Ar ages of bronzite,
enstatite, pigeonite, carbonaceous and nonclassified chondrites

The K-Ar age can be calculated from the K^{40} and Ar^{40} content
according to the equation:

$$t = \frac{1}{\lambda} \ln \left(1 + \frac{A^{40}}{K^{40}} \frac{1+R}{R} \right)$$

λ is the total decay constant and R is the branching ratio of the decay
constant for electron capture and β^- decay. The values used for λ and R
are $5.32 \times 10^{-10} y^{-1}$ and 0.123, respectively.

Fig. 2 shows a histogram of the K-Ar ages of stone meteorites. Most of
the meteorites have ages between 3.5 and 4.5 b.y. Few have ages higher
than 4.5 b.y. the highest one is that of Norton County with 5.1 b.y.
In most cases the error in the measurements is not larger than 4 to 6%,
in some cases even smaller. Different authors agree, that meteorites with

smaller ages suffered gas losses. Meteorites with ages below 0.25 b.y. have not been found.

In Fig. 3 and 4 the K-Ar ages of chondrites have been separately plotted for the hypersthene chondrites (including the amphoteritic chondrites) and for the bronzite-, enstatite-, pigeonite- and carboneous chondrites. This division corresponds to the classification of UREY and CRAIG (1953), who distinguish between a low-iron and high-rion group with a total iron content of 22% and 28% respectively. KEIL (1964) and EBERHARDT and GEISS (1964) (see also ANDERS, 1964) argued with somewhat limited statistics, that there exists a difference in the K-Ar age distributions between both groups. With a larger number of data (ZÄHRINGER, 1964; KAISER and ZÄHRINGER, 1965) it can now safely be concluded, that the K-Ar age distributions for hypersthene chondrites and for bronzite chondrites exhibit a difference. A great number of hypersthene chondrites lie below 2 b.y. In this histogram 9 black chondrites (HEYMANN, 1965a) have been included. They fit into this general pattern, although they may have lost more gases than the normal hypersthene chondrites, because they may have suffered stronger heating due to their darker colour (HEYMANN, 1965a). The amphoteritic chondrites show a very interesting distribution. Though they are a subclass of the hypersthene chondrites their age distribution resembles more that of the bronzite chondrites. Most of them (16 out of 19) have ages between 3.5 and 4.5 b.y. Recent attempts to extend the K-Ar age measurements to separated mineral fractions of a meteorite, have shown, that one might find different ages (AUER et al., 1965; SCHAEFFER et al., 1965, MEGRUE and STOENNER, 1965). But one should, however, be careful in interpreting these results in terms of different solidification times of the various mineral components, as diffusion might have influenced them to a different extent.

5. Argon losses and comparision with the He⁴-ages

The U, Th-He⁴ ages are gas retention ages as well as the K-Ar ages. Due to the faster diffusion of He through the silicate lattice as compared to Ar^{40}, these ages reflect even more sensitive the temperature history of meteorites. An experimental disadvantage of the U-He method is the complicated analytical technique needed to determine the uranium and thorium content. HAMAGUCHI et al. (1957) succeeded in measuring the uranium concentration by neutron activation technique. The U content in chondrites ranges between 10 and 15 ppb. Achondrites show large variations from 10 to 120 ppb (REED and TURKEVICH, 1957; KÖNIG and WÄNKE, 1959; GOLES and ANDERS, 1962). A Th/U ratio of 3.6 has been found, which is similar to the terrestrial one (BATE et al., 1959).

More recently LOVERING and MORGAN (1964) showed, that this ratio may vary up to 5.0 according to the content of uranium. But in order to calculate He⁴ ages for chondrites, one may use average values for the uranium content and the Th/U ratio. The measured He⁴ has to be corrected for a minor contribution from spallation helium (see Table 1). In Fig. 5 and 6 the He⁴ ages have been plotted for hypersthene- and amphoteritic chondrites and for bronzite-, enstatite- and pigeonite-chondrites respectively. A comparison with the corresponding K-Ar

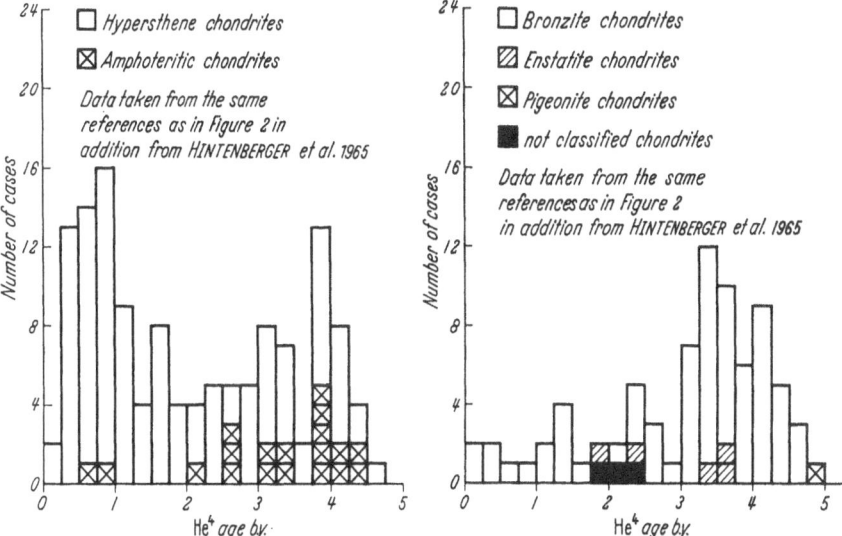

Fig. 5. Histogram of He⁴ ages of hypersthene and amphoteritic chondrites

Fig. 6. Histogram of He⁴ ages of bronzite, enstatite, pigeonite and nonclassified chondrites

age distributions shows the similarity of both, though the He⁴ age distribution is shifted to somewhat lower values. This fact can be explained by the faster diffusion of He⁴. The difference between hypersthene- and bronzite chondrites is also reflected in the He⁴ age histogram (HINTEN-BERGER et al., 1964; ZÄHRINGER, 1964; VINOGRADOV and ZADOROZHNY, 1964; HINTENBERGER et al., 1965). There are much more hypersthene chondrites than bronzite chondrites with He⁴ ages below 2 b.y. This difference is even more pronounced than for the K-Ar ages (see Fig. 7 and 8). There are 3 groups of meteorites (KIRSTEN et al., 1963; ANDERS, 1964): Meteorites with high and consistent K-Ar and He⁴ ages, meteorites with low and consistent and meteorites with inconsistent ages. For the latter ones the He⁴ ages are in most cases smaller than the K-Ar ages. Hypersthene chondrites are represented in all three groups, whereas bronzite chondrites appear only in the first and third group. In order to explain the gas retention ages four processes were discussed, which could have been responsible for gas losses: 1. cooling down of the parent-body after melting, 2. diffusion losses at elevated temperatures, when the meteorites are still incorporated into the parent-body, 3. solar heating, when the meteorite is on an orbit approaching the sun (Goles et al., 1960) and 4. gas losses during collisional break-up (Kirsten

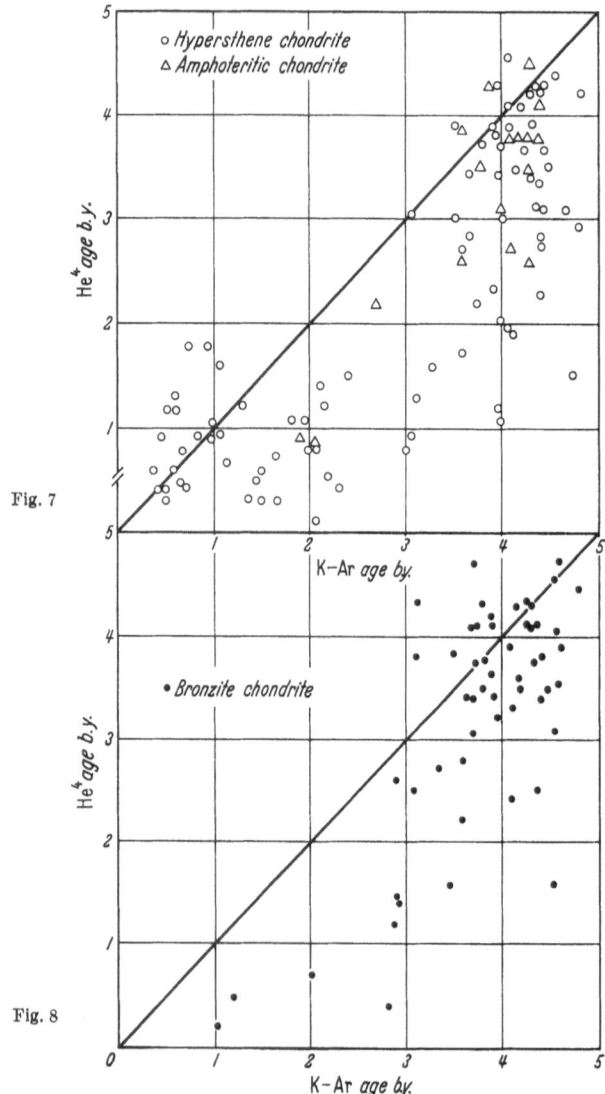

Fig. 7

Fig. 8

Fig. 7. Comparison of K-Ar and He⁴ ages of hypersthene and amphoteritic chondrites
Fig. 8. Comparison of K-Ar and He⁴ ages of bronzite chondrites

et al., 1963; Fechtig et al., 1963; Eberhardt and Geiss, 1964). Diffusion losses of rare gases in meteorites have been considered previously (Geiss and Hess, 1958; Stoenner and Zähringer, 1958; Eberhardt and Hess, 1960). Fechtig et al. (1963) carried out experiments, to study the diffusion of rare gases in meteorites. They found activation energies for argon of about 45 Kcal/mol and such diffusion constants, that even at temperatures up to 50° C no significant diffusion losses of neither argon nor helium should occur. They showed, that diffusion in the meteorites strongly depends on the grain size distribution. Small grains lose their rare gase at low temperatures. At higher temperatures only the larger grains contribute to the rare gase relase. Meteorites with the same thermal history may have suffered different gas losses, due to different grain size distributions. An extensive discussion of these subjects is given by FECHTIG and KALBITZER in this book.

The consistent ages for the hypersthene and for the bronzite chondrites can be explained by complete degassing about 4.5 b.y. ago. The time for cooling to temperatures where helium and argon are retained, must have been short (10^8 y.), if these gas losses occured during cooling after melting of the meteorite parent-body. Otherwise the K-Ar and He^4 ages should not be consistent. ANDERS (1964) used the results of the diffusion experiments by FECHTIG et al. (1963) and estimated the size of parent-bodies of some meteorites. The results show, that a radius of less than 200 km is compatible with the diffusion data. Another method to estimate this size is provided by the decay of J^{129} into Xe^{129} with 17 m.y. half life. J^{129} seems to have been present during condensation and accumulation into larger bodies. A fraction of the decay product Xe^{129} could have been retained. This Xe^{129} causes an isotopic anomaly in the meteoritic xenon. REYNOLDS (1960d) detected in the Richardton chondrite a 50% excess of Xe^{129}. Shortly after this discovery a Xe^{129} anomaly was also reported for many other meteorites (ZÄHRINGER and GENTNER, 1960; REYNOLDS, 1960a, c; ZÄHRINGER, 1962). Although there is still no general agreement about the process, that produced the J^{129}, one can estimate under certain assumptions a formation interval, which is the time between cessation of the J^{129} production and beginning of Xe^{129} retention in meteoritic material. In this way one arrives again at times in the order of 10^8 y. This gives an upper limit of 200 km diameter for the size of the parent body (ANDERS, 1964).

Meteorites with consistent high K-Ar and U-He-ages might have originated from the surface regions of a larger parent body. The surface layers of larger bodies have cooled down fast to temperature where gases were retained. Meteorites could have been stored there without gas losses for long times.

To explain an origin from the inner parts of a large body, one has to assume, that about 4.5 b.y. ago fast cooling occured by a break-up of the parent body. ANDERS (1964) mentioned, that in order to explain the scatter from 3.5 to 4.5 b.y. one must assume, that a series of break-ups took place.

In Fig. 7 the amphoteritic chondrites exhibit a remarkable behaviour. In contrast to the other hypersthene chondrites most of them have high consistent K-Ar- and U-He-ages. Thus they are more similar to the bronzite chondrites (HEYMANN, 1965), as can be seen from Fig. 8. On the other hand their distribution of the exposure ages is similar to hypersthene chondrites. They probably came from the same parent body, but originated as small bodies by several events.

The meteorites with inconsistent ages, mainly hypersthene chondrites, can derive from process 2 to 4. Some meteorites (10%) exhibit gas losses, which have occured during their exposure to the cosmic radiation

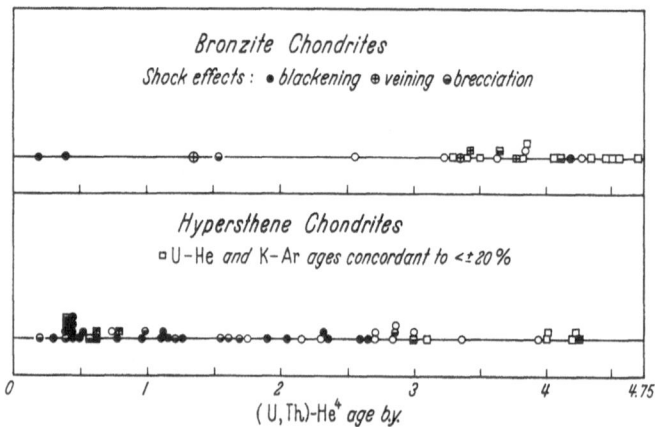

Fig. 9. Correlation between shock effects and He⁴ gas retention ages (reproduced from ANDERS, 1964)

(KIRSTEN et al,. 1963; HINTENBERGER et al., 1964; EBERHARDT et al. 1965). This can be seen from the low He^3/Ne^{21} ratio (<4) together with low rare gas contents. They probably suffered gas losses by solar heating, while they were on orbits close to the sun.

Preferential losses of helium during storage at higher temperatures in the parent-body could also be responsible for the inconsistent ages.

Another possible process is incomplete degassing by collisional break-up. Many meteorites exhibit certain symptoms, which are supposed to be produced by shock. These are blackening, veining and brecciation (FREDRIKSSON et al., 1963). Fig. 9 shows a compilation of shock data after PRIOR and HEY (1953), KEIL (1960) and ANDERS (1964) as reproduced from ANDERS (1964). Shock effects, if correctly defined, seem to be frequent among meteorites with low ages. There is a difference between hypersthene- and bronzite chondrites, as in the case of age distributions. Shock effects among the latter seem to be also rarer.

The hypersthene chondrites show in addition a grouping around 0.5 b.y. Meteorites of this group are thought to be products of one event (KIRSTEN et al., 1963; ANDERS, 1964; ZÄHRINGER, 1964). They must have been completely degassed. Incomplete degassing would result in differences between the K-Ar and U-He ages, because of the preferential escape of He⁴. One believes for this reason, that this group has a common origin, perhaps as the result of a collision 0.5 b.y. ago.

6. Comparison with Rb-Sr and Pb-Pb ages

Besides the K-Ar and U, Th-He method the decay of Rb^{87} to Sr^{87} and the decay of U^{235} to Pb^{207} and U^{238} to Pb^{206} have been used to date stone meteorites. The modes of decay and the half lives are as follows:

$Rb^{87}(\beta^-, 5.0 \times 10^{10}y.) Sr^{87}$

$Th^{232}(\alpha, \beta^-, 1.39 \times 10^{10}y.) Pb^{208} + 6 He^4$

$U^{235}(\alpha, \beta^-, 7.1 \times 10^8 y.) Pb^{207} + 7 He^4$

$U^{238}(\alpha, \beta^-, 4.51 \times 10^9 y.) Pb^{206} + 8 He^4$

$K^{40}(K, \beta^- 1.30 \times 10^9 y.) Ar^{40}, Ca^{40}$

The K-Ar and the U, Th-He methods on principle date the time elapsed, since meteoritic material has cooled down to temperatures,

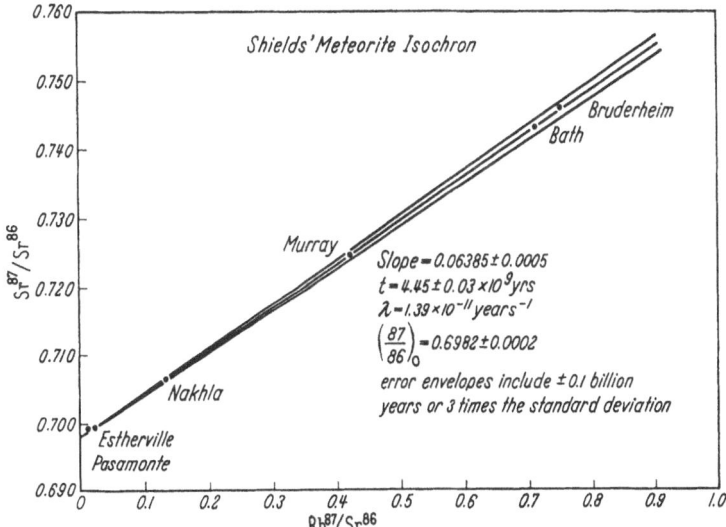

Fig. 10. SHIELDS' Rb-Sr meteorite isochrone (reproduced from SHIELDS, 1964)

where Ar^{40} and He^4 was retained (gas retention ages). The Rb-Sr and the Pb-Pb method on the other hand date the cessation of chemical differentiation in the Rb-Sr and U, Th-Pb system (solidification ages). Since the separation of solid elements occured at higher temperatures than gas retention, radioactive clocks with the nonvolatile Sr and Pb are expected to have started somewhat earlier.

A difficulty inherent in the Rb-Sr and Pb-Pb method is, that seperation of mother- and daughter element has been incomplete. This results in two unknown quantities in the age equation: solidification age and primordial isotopic composition of Sr respectively of Pb. A standard approach to overcome this difficulty has been constructing isochrone diagrams, which implies a number of specific assumptions concerning the origin of the meteorites. These problems are discussed in detail in the references given below for work on the Rb-Sr and Pb-Pb method.

Rb-Sr

The first measurements of Rb-Sr age of stone meteorites have been performed by HERZOG and PINSON (1956) and by SCHUMACHER (1956a, b, c, d). Later work added an amount of new data and improved to a large extent the experimental technique (WEBSTER et al., 1957; GAST, 1960, 1961, 1962, 1963; PINSON et al., 1962, 1963, 1964; MURTHY and Compston, 1964; BEISER and PINSON, 1964; SHIELDS, 1964). Most of the earlier studies suffered from poor analytical precision and show for this reason in a number of cases discrepancies in the analytical results. The analytical technique of the method has been much improved now and results with high precision can be obtained.

Fig. 10, which has been reproduced from SHIELDS (1964), shows an isochrone for 6 meteorites each belonging to a different class. Although the precision of these measurements is very high, one has to remember, when interpreting the results in terms of a solidification age, that errors are introduced by the uncertainty of the half live of Rb^{87}. There exists still a disagreement of the measurements based on direct physical methods (see for example BEARD and KELLY, 1961; LEUTZ et al., 1962). Setting aside these difficulties, the Rb-Sr results seem to indicate, that 4.5 b.y. ago the seperation of Rb from Sr took place in meteoric material.

Pb-Pb

The decay of uranium into lead deserves special attention. As the two isotopes U^{235} and U^{238} decay into different lead isotopes Pb^{207} and Pb^{206} one can combine both age equations and eleminate the uranium content in the resulting equation. The Pb-Pb method is therefore based only on isotope ratio measurements, which is a great advantage over the Rb-Sr method. In addition the involved half lives are accurately known. On the other hand the contamination problem is more serious and has by no means been solved as yet. The results show, that terrestrial lead has been for the same time in a similar uranium environment as the meteoritic lead and can therefore hardly be distinguished from the latter. Another startling point is, that in most meteorites the uranium content cannot account for the observed lead values (HAMAGUCHI et al., 1957). No explanation for this inconsistency has been found. An indication that weathered samples are contaminated with terrestrial lead has been reported by STARIK et al. (1959, 1960) for iron meteorites.

After PATTERSON (1955, 1956) various Pb-Pb ages of meteorites have been reported (HESS and MARSHALL, 1960; STARIK et al., 1959, 1960; MURTHY and PATTERSON 1962; MARSHALL, 1962).

For solving the Pb-Pb age equation again the isochrone method is used. Fig. 11, reproduced from ANDERS (1963), summarizes the Pb-Pb data.

This isochrone seems to indicate, that seperation of U and Th from Pb in meteoritic material took place 4.6 b.y. ago. But at present one hesitates to draw final conslusions from the Pb-Pb results as far as the history of meteorites is concerned.

Comparing the results of the three methods K-Ar, Rb-Sr and Pb-Pb one has to remember, that they date different events. As the K-Ar

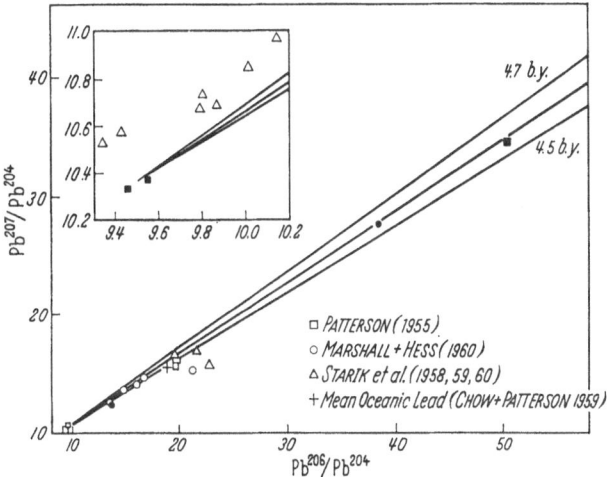

Fig. 11. Pb-Pb meteorite isochrone (reproduced from ANDERS, 1963)

ages are often lowered by gas losses during the further history of the meteorite, one has to compare the upper limit of the K-Ar age distribution with the solidification ages. Taking 4.5 b.y. as the time where chemical differentiation in meteoritic material ceased, one must conclude, that only very shortly afterwards the material had cooled down to temperatures, where argon was retained. Besides some other meteorites the achondrite Norton County with a K-Ar age of 5.1 b.y. does not fit into this picture. Up till now there seems to be no process other than radioactive decay of primordial K^{40}, which could account for the high Ar^{40} value.

7. Break-up and cosmic ray-exposure ages

The parent bodies of the meteorites have been broken up at some time by collision or by some other unknown processes. As long as the meteorite specimens were incorporated in a larger body, they were shielded from cosmic ray bombarment. When the sample came to the surface within the penetration depth of cosmic rays, about one meter, new products from nuclar reactions were induced and isotopic variations caused. Under certain assumptions one can calculate the time of irradiation (radiation age or cosmic ray-exposure age) by determining the present-day amounts of cosmic ray produced stable and radioactive isotopes (ARNOLD, 1961; SCHAEFFER, 1962; GEISS et al., 1962; HONDA and

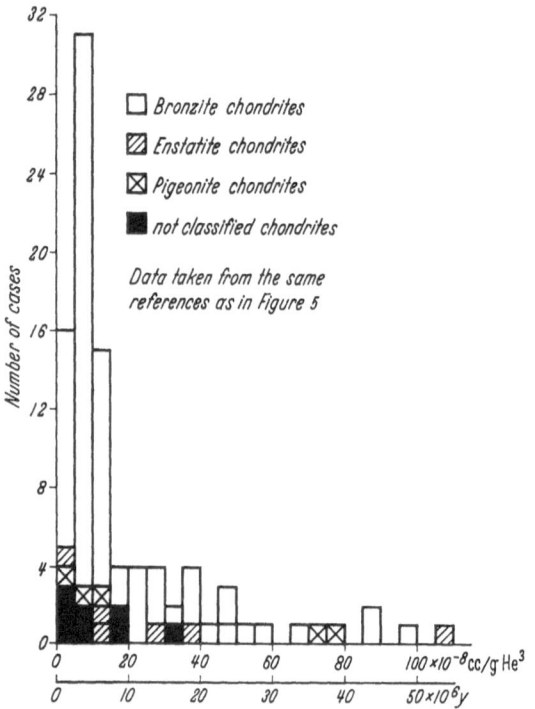

Fig. 12. Histogram of the exposure ages of bronzite and some other chondrites

ARNOLD, 1964). Though a number of such pairs have been used, most radiation age data now available for stone meteorites are T-He^3 radiation ages. These are calculated assuming a He^3 production rate of about 2×10^{-8} cc STP per gram per million years. This assumption seems to be justified as it has been observed, that the tritium content of stone meteorite is fairly constant (c.f. GEISS et al., 1960; GOEBEL and SCHMIDLIN, 1960; BEGEMANN, et al., 1957, 1959). In Fig. 12 and 13 the He^3 radiation ages of bronzite and hypersthene chondrites are compared. It is obvious, that bronzite chondrites exhibit a behaviour different to that of the hypersthene chondrites. Most of the bronzite chondrites have radiation ages of about 4 m.y., whereas the hypersthene chondrites show no clustering. Now with much more data the suggestion of such a clustering by GEISS et al. (1960) can be confirmed, however, rather for the bronzite than for the hypersthene chondrites. The number of enstatite chondrites, pigeonites and amphoterites is too small yet to exhibit such correlation. Achondrites have been excluded from the figure as their number is also limited, but it should be mentioned, that aubrites seem to have radiation ages higher than chondrites (KIRSTEN et al., 1963; EBERHARDT and GEISS, 1964; EBERHARDT et al., 1965). Six out of nine lie around 40 m.y.

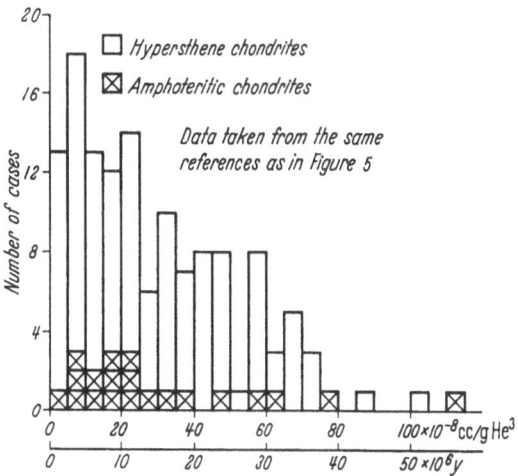

Fig. 13. Histogram of the exposure ages of hypersthene and amphoteritic chondrites

The fact, that most bronzite chondrites are concentrated around 4 m.y. points to one major event as the result of which the meteorite fragments were formed. Specimens with higher ages may be remnants of previous events. It seems, that subsequent collisions had not occured, because of the narrow clustering around 4 m.y. The mean time between two collisions must be smaller than 4 m.y. in this case. Recent calculations on capture life times of objects from the asteroidal belt and from lunar ejecta show, that the stony meteorites must have originated close to the earth (ARNOLD, 1965). A lunar origin seems to be favoured, especially for the bronzite chondrites. Hypersthene chondrites differ in their history from bronzite chondrites. Their radiation ages show a wide and almost continuous distribution. Like bronzite chondrites they exhibit no influence from the last collision on their gas retention age. This means, that larger gas losses mainly observed in hypersthene chondrites must have occured while these specimens belonged to larger bodies. The low-age group of hypersthene chondrites clearly shown in Fig. 3 and 5 probably been created in one single event 0.5 b.y. ago. The continuous distribution of radiaton ages of hypersthene chondrites can be explained by multiple collisions. Experimental evidence for this is given by the

New Concord meteorite with two exposure ages of 2 and 36 m.y. (Zähringer, 1964). From the estimated size of this meteorite one can conclude, that these two ages cannot be explained as depth effect of cosmic
ray penetration.

B. Iron Meteorites

Introduction

The radiogenic ages of stone meteorites, as determined by several
independent methods, agree in the maximum value of 4.5–5 b.y.
with the age of the earth, as deduced by the Pb-Pb method. This result
is no longer surprizing. Recently some evidence has been accumulated,
that stony meteorites might have originated close to the earth and
as such have an evolutionary history similar to terrestrial matter.
This evidence is derived from their cosmic ray exposure ages, which are
a few million years and are relatively low. Bronzite chondrites even
have a pronounced peak at 4 m.y. and according to calculations by
Arnold (1965) the moon is a favorite candidate as parent body. The
cosmic ray exposure ages of iron meteorites are about a hundred times
higher than those of chondrites, a fact well established on about 200
meteorite samples. This distinct difference also points to a difference in
the history and place of origin of these two types of meteorites.
It is more likely that iron meteorites have their origin in the asteroidal
belt. The crucial question arises: What are the radiogenic ages of iron
meteorites? Attempts in this direction were made a long time ago by
Paneth and coworkers, who tried to determine He^4-ages of iron meteorites. It turned out, however, that the He found is mainly produced in
spallation reactions by cosmic rays. In addition, the measured U concentrations were too high by several orders of magnitude, as shown by
Reed and Turkevich (1957). In fact the U content is less than $10^{-10}g/g$
and it would be extremely difficult to obtain He^4 ages.

The lead isotope determinations in troilite inclusions of iron
meteorites gave the lowest abundances for the radiogenic isotopes
compared to Pb^{204}. This lead is considered to be primordial lead. Although
also some lead enriched in the radiogenic isotopes has been found in
iron meteorites, an age determination with the Pb-Pb method has not
been possible so far.

The Re-Os-method has also been applied to iron meteorites. Unfortunately the Re^{187} half life is very difficult to determine by physical
methods, because of its low decay energy (< 8 keV). In addition the
chemical separation of rhenium and osmium is very poor in iron meteorites. Though the rhenium content shows large variations from 5 ppb
to 4.8 ppm, the Re^{187}/Os^{186} ratios lie within the narrow range from 2.5 to 8.

The Re-Os age thus derived lies between 4.5 and 5 b.y., but it needs further investigations (Herr et al., 1961).

A possibility to determine the cooling time of iron meteorites is given by the K-Ar-method. The abundance of K can be expected to be 10^4 to 10^5 times higher than that of U and the spallation produced Ar^{40} is about 500 times smaller than the corresponding amount of He^4. The diffusion of rare gases in metals is also very low, so that a K-Ar-age could represent a reliable age for the formation of their parent bodies. No further assumption than complete degassing at the time of solidification is required.

Stoenner and Zähringer (1958) have tried to apply the K-A-method to iron meteorites by means of n-activation analysis. The ages obtained have been high and ranged from 5 to 13 b.y. The very high ones have been obtained from samples with extremely low K content and deserve therefore less consideration. In principle this work showed that K-Ar ages of iron meteorites are feasible. More accurate values can be obtained, when further investigations on spallation corrections and analytical improvements are made.

Fisher (1963, 1965) has added a few more results, using the same experimental procedures. He obtained similar K/Ar-values with the exception of Sikhote Alin where a much higher K-content was determined.

Recently Müller and Zähringer (1966) re-investigated the whole problem on a broad basis, using different techniques and a larger variety of samples. These experiments are still being continued and the following report gives the present state of progress.

1. Experimental

The general information about n-activation is given in chapters 1 and 2, so that only some details pertaining to this special problem have to be added.

The Ar^{40} content in iron meteorites consists of at least three components: spallation produced Ar^{40}_{sp}, possible atmospheric Ar^{40} and radiogenic Ar^{40}_{rad}.

The spallation Ar^{40}_{sp} itself consists of Ar^{40} directly produced and of Ar^{40} resulting from the decay of spallation produced K^{40}. The sum of both Ar^{40}_{sp} has been determined in all samples experimentally by mass spectrometry. First the Ar^{40}/Ar^{38} ratio was determined with a specially developed degassing technique (Lämmerzahl, 1966). Thus minimum values of the Ar^{40}/Ar^{38} ratios could be achieved, which range from 0.17−0.22, depending on the cosmic ray exposure ages. These Ar^{40}/Ar^{38} measurements are equivalent to the K^{40}/K^{41} determination by Voshage and Hintenberger and give in addition to spallation correction a new method for cosmic ray exposure ages. In an adjacent sample the total amount of Ar^{38} and Ar^{36} was determined by conventional mass spectrometric analysis, so that the Ar^{40} is now known within a few percent for each meteorite investigated.

In order to keep the atmospheric contamination low, the samples were sealed in quartz tubes a few days before irradiation and predegassed at 100° C in vacuum. The extraction of the A^{41} after the irradiation occured in steps. Usually three fractions during 10 min degassing at 400° C, 800° C, and 20 min at 1800° C were analyzed. Carrier argon was added, the gas purified and filled into a Geiger counter with argon-ethylacetate filling. Measurements with low Ar^{41} content in the first two fractions are considered more reliable. The Ar^{41} of the low temperature degassing was considered to be atmospheric contamination. The error given for the Ar^{40} content is estimated from this low temperature gas release. In addition many specimens have partially been dissolved in HCl after predegassing at 400° C and the Ar and K were measured in the fractions. Since inclusions are more easily dissolved than the pure metal phase, the first fraction contained a higher K concentration. By this technique it should be proved whether the Ar^{40} content is proportional to the K or not. Besides Ar^{41} with a half life of 1.8 hrs also Ar^{37} with 34 days half life could be determined from the decay curve. Usually the Ar^{37} content is in good agreement with the spallation produced Ar^{36} content (1 cpm/g Ar^{37} corresponds to 1.3×10^{-8} cc/g Ar^{36}). In samples with high K content, however, Ca could also be seen from the Ca^{40} (n, α) Ar^{37} reaction (1 cpm/g Ar^{37} corresponds to 80 ppb Ca). In general the Ca and K contents are nearly equal in the inclusions of the iron meteorites investigated. A further small contribution to Ar^{41} comes from K^{41} (n, p) Ar^{41} reaction, where 1 μg K yields 31 cpm. All side reactions were calibrated and the results were corrected accordingly.

The calibration of the Ar isotopes has been performed together with K standards. A quartz tube containing about 100 mm³ air was irradiated together with a sample. By volume dilution, a fraction of 1 percent was used for the Ar^{41} activity, while the total amount was used for the Ar^{37} calibration. This calibration was performed inside and outside of meteorite samples and an average screening effect of 15% was determined for a 10—15 g sample. The routine control of the flux was taken from a K standard. The flux varied less than 15% between various runs in a 2 year period.

The experimental procedure for the K determination has already been described in section III. It has been emphasized, that this technique is especially powerful in detecting small amounts of K. The sensitivity is in the order of 10^{-11} g. Potassium has been found to be very inhomogeneously distributed in all meteorites. It is mainly located in silicate, schreibersite and troilite inclusions.

The Ar and K therefore have to be determined both in the same specimen. In many cases 10 to 20% of the sample were dissolved in a HCl solution. K and Ar have been analyzed separately in the leach fraction and in the remaining sample. In this way it could be seen, whether the Ar⁴¹ content is proportional to the K content and therefore of radiogenic origin, or whether the age of the inclusions is different from the age of the remaining sample.

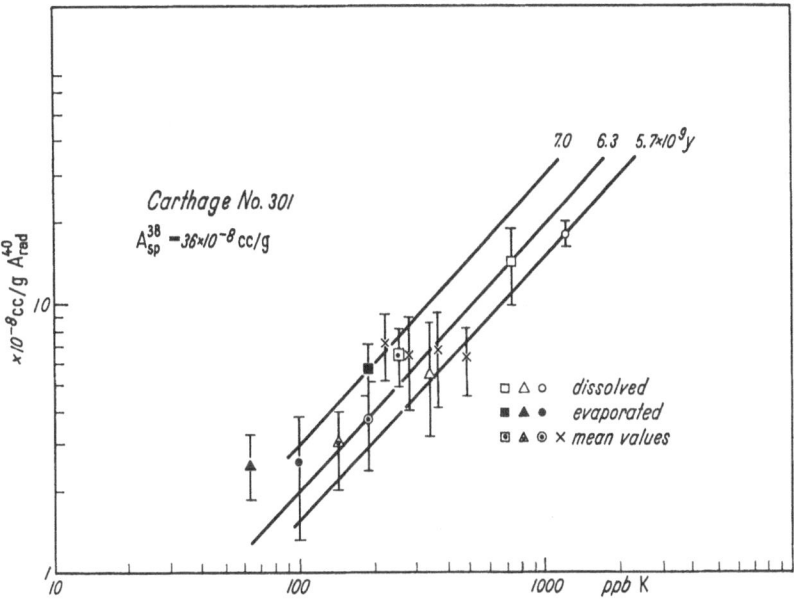

Fig. 14. Radiogenic Ar⁴⁰ and K content of the Carthage meteorite

From these etching experiments it also could be learned, that no etching should be applied to remove surface contamination. In those specimens, where the inclusions consist of narrow schreibersite lamellae, the solution with about 20% of the sample weight contains up to 90% of the total K content. In all previous work this mistake was made. In fact it would not affect the age, if K and Ar would be located at the same places. Unfortunately, K and Ar seem to have had differing diffusion during the meteorite's history as one might expect. This problem, however, will be discussed later.

2. Results

In Table 2 the detailed results of the Carthage meteorite are given. This meteorite has a high spallation content of 36×10^{-8} cc/g Ar³⁸. Accordingly the correction for Ar_{sp}^{40} and K_{sp}^{41} has been applied. The whole

Table 2. *Detailed results of argon and potassium analysis of the Charthage meteorite*

Name	Weight in g	A^{41} dpm/g	$A^{40} \times 10^{-8}$ cc/g	$A^{40}_{rad} \times 10^{-8}$ cc/g	K_{prim} 10^{-9} g/g	A^{40}_{rad}/K_{prim} cc/g	Age 10^9 y
Carthage 301, Om							
$A^{38}_{sp} = 36 \times 10^{-8}$ cc/g	11.4	580± 90	13.9 ± 2.2	6.4 ± 2.5	280	0.23±0.09	6.5 + 0.6 − 0.9
$A^{40}_{sp} = 7.5 \times 10^{-8}$ cc/g	9.7	626 ± 60	13.8 ± 1.5	6.3 ± 1.8	480	0.13±0.04	5.5 + 0.5 − 0.6
	8.3	615 ± 70	14.7 ± 1.7	7.2 ± 2.0	225	0.32±0.09	7.1 + 0.4 − 0.6
$K^{41}_{sp} = 1.1 \times 10^{-9}$ g/g	8.3	590 ± 90	14.1 ± 2.2	6.6 ± 2.5	370	0.18±0.07	6.1 + 0.6 − 0.9
$A^{37}_{sp} = 17$ dpm/g	2.0 d	525 ±100	13.1 ± 2.5	5.6 ± 2.8	373	0.15±0.07	5.7 + 0.7 − 0.1
	5.5 e	400 ± 20	10.0 ± 0.5	2.5 ± 0.8	65	0.38±0.12	7.4 + 0.5 − 0.7
	7.5 a	420 ± 28	10.5 ± 0.7	3.0 ± 1.0	147	0.20±0.07	6.3 + 0.5 − 0.8
	0.90 d	840 ±170	21.8 ± 4.4	14.3 ± 4.7	734	0.19±0.06	6.2 + 0.5 − 0.7
	5.50 e	510 ± 30	13.2 ± 0.8	5.7 ± 1.1	184	0.31±0.06	7.1 + 0.3 − 0.4
	6.40 a	537 ± 50	13.9 ± 1.3	6.4 ± 1.6	262	0.24±0.06	6.6 + 0.4 − 0.5
	0.51 d	980 ± 50	25.4 ± 1.3	17.9 ± 1.6	1260	0.14±0.01	5.6 + 0.1 − 0.2
	6.11 e	390 ± 40	10.1 ± 1.0	2.6 ± 1.3	100	0.26±0.13	6.7 + 0.9 − 1.2
	6.62 a	437 ± 40	11.3 ± 1.1	3.8 ± 1.4	190	0.20±0.07	6.2 + 0.6 − 0.7

d = dissolved
e = evaporated
a = average

specimen has numerous small schreibersite veins, which account for the relative high K content of 280 ppb on the average. The results of the first 4 samples were obtained by total melting. The next 3 samples were partially dissolved. It can be seen, that quite differing K concentrations have been achieved and in general the Ar follows the K content (Fig. 14). The inclusions, however, have somewhat lower Ar/K values. This observation can be explained as follows: Either the inclusions have partially lost their Ar^{40} due to a higher diffusion constant or K diffused over a long period of time from the iron phase into the inclusions.

The authors favor the first interpretation, since the Ar/K values both in the iron phase and inclusions are independent of the K content. The difference seems to be due more to the mineralogical structures. In this case the age of the iron phase gives the most likely time, since the cooling of the meteorite parent body started. In the second case the mean value of both fractions would give the real age. In the case of Carthage the average age of the inclusions is 5.8 b.y., while the average

of all samples would be 6.3 b.y. The Treysa meteorite exhibits much
larger discrepancies (see Fig. 15). It contains large lamellae of schreibersite with silicate inclusions. Here the inclusions have ages of only 2.7 b.y.,
while the iron phase ages are more close to 6 b.y. Again the size or the K
concentration has no affect on the age. The studies of a polished section
shows kataclastic effects. In a similar way several other meteorites
have been investigated as listed in Table 3 and Fig. 16. The values given
are mean values of at least 5 to 10 independent runs. Two Canyon Diablo

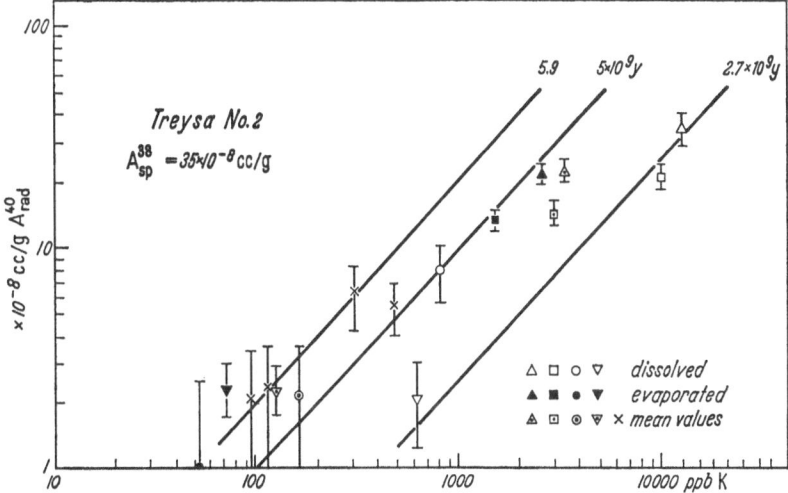

Fig. 15. Radiogenic Ar⁴⁰ and K content of the Treysa meteorite

samples with different spallation content of 0.51 and 10×10^{-8} cc/g Ar³⁸
for number 3 and 266 resp. were analysed. The ages of 8.1 and 7.4 b.y.
are quite high. Perhaps Canyon Diablo 3 with the low K concentration
of only 42 ppb has some small amounts of inherited Ar⁴⁰, so that it
only represents an upper limit. The Canyon Diablo 266 sample has
numerous schreibersite inclusions and a five time higher K content.
It's age of 7.4 b.y. is derived from many samples with differing K concentrations.

The Ar/K ratios of the Charcas samples showed large variations, so
that the average age has a rather large error. According to mineralogical
investigations many troilite inclusions are re-crystallized in a very
strange way and are randomly distributed. The scattering K/Ar values
would be in agreement with this mineralogical observation. Gundaring
has a relatively high and homogeneous K and Ar content. Unfortunately,
not too many specimens of Gundaring have so far been available.

The Sikhote Alin samples 309 and 326 have differing spallation Ar³⁸
content of 3.9 and 9.5×10^{-8} cc/g respectively. Both samples are highly

kataclastically deformed 309 more than 326. The Ar/K value of the schreibersite inclusions in 309 E is also extremely low. This meteorite certainly suffered from shocking some portions to a larger and some to

Table 3. *Mean $\overline{K}$ and radiogenic $\overline{Ar}^{40}$ values of iron meteorites with large primordial potassium content*

Meteorite	$\overline{A^{40}_{rad}}$ $\times 10^{-8}$ cc/g	$\overline{K}_{prim}$ 10^{-9} g/g	$\overline{A^{40}_{rad}}/\overline{K}_{prim}$ cc/g	Age 10^9 y
Canyon Diablo 3	2.3 ± 1.0	42	0.55 ± 0.10	$\leqq 8.1 + 0.3$ -0.4
Canyon Diablo 266	7.7 ± 2.0	203	0.38 ± 0.11	$\leqq 7.4 + 0.5$ -0.6
Carthage	5.7 ± 1.8	279	0.20 ± 0.02	6.3 ± 0.2
Carthage I	12.6 ± 3	790	0.16 ± 0.03	5.8 ± 0.3
Charcas	1.6 ± 0.9	94	0.17 ± 0.10	$5.9 + 0.9$ -1.4
Gundaring	3.6 ± 1.2	184	0.19 ± 0.03	$6.2 + 0.3$ -0.4
Sikhote Alin 309	2.5 ± 1.0	105	0.24 ± 0.02	6.6 ± 0.2
Sikhote Alin 309 I	3.1 ± 0.5	1840	0.017 ± 0.003	2.2 ± 0.3
Sikhote Alin 326	2.0 ± 0.7	49	0.41 ± 0.04	$\leqq 7.5 \pm 0.2$
Treysa	3.4 ± 1	215	0.16 ± 0.02	$5.9 + 0.2$ -0.3
Treysa I	16 ± 2.6	6000	0.026 ± 0.005	2.8 ± 0.3

(I = inclusions)

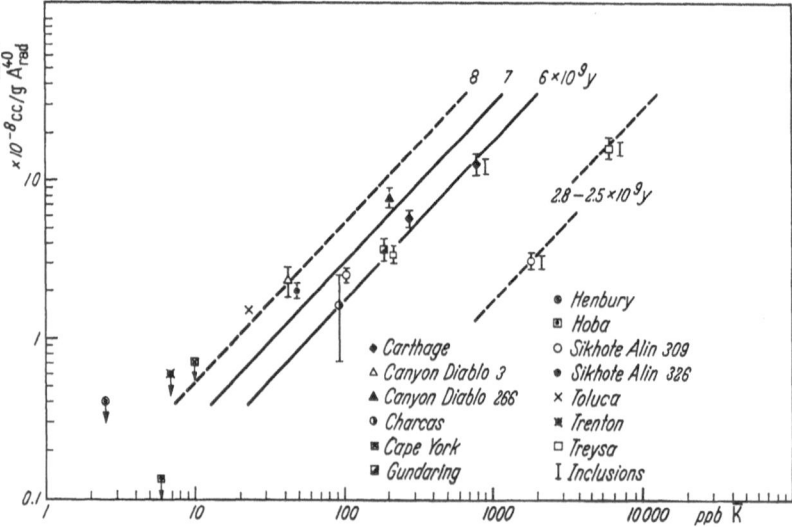

Fig. 16. Mean values of radiogenic $\overline{Ar}^{40}$ and of $\overline{K}$ in several iron meteorites

a lesser degree. This event may have occurred in recent times, since it was observed, that some inclusions also lost some of their Ar³⁶. A summary of the results is given in Fig. 16 and Table 3. Also measurements of further meteorites are included, whose K and Ar concentration is very low. Hoba is such a case. It is an ataxite and its K-Ar-age is very low. Cape York, Toluca, Henbury, and Trenton have very low K and Ar concentrations. The corrections for all these cases, however, are too large so that no reliable age can be given. These measurements are very important for the interpretation of the other results. From samples with high Ar^{38}_{sp} content (Henbury 14.3, Trenton 22, Hoba 4.1×10^{-8} cc/g Ar^{38}_{sp}) and low primordial K content one learns that the corrections for spallation products are correct. Samples with low Ar^{38}_{sp} and low primordial K content prove, that the amount of inherited or atmospheric Ar^{40} as well as K contaminations can be lower than a few 10^{-9} cc/g Ar^{40} or some ppb K respectively. This means, that it is most likely that samples with higher primordial K content are not affected by above contaminants.

3. Summary

This investigation showed, that single specimens from the same meteorite with differing K content give, with the exception of inclusions, fairly constant Ar/K ratios. Isochrones can be drawn and for the examples Carthage an average age of 6.3 b.y., for Treysa 5.9 b.y. are obtained. Similar results with ages up to 7.5 b.y. have been obtained for other meteorites, for which the details are given in MÜLLER and ZÄHRINGER (1966).

Furthermore one can use the average $\overline{Ar^{40}}$ and $\overline{K}$ values for different meteorites and as seen in Fig. 16 the points are, except for the inclusions, all between the 5 and 7.5 b.y. isochrones. The fact that all the meteorites and possibly even different specimens do not have a single age is probably due to their complex history. They also might be derived from several parent bodies. Hoba is an ataxite and its young age might be connected with its mineralogical structure. The number of cases is too small to see already some systematics.

The complex history can be seen in a microscopic picture. Irregular inclusion of troilite, schreibersite, silicates etc. and strange combinations of such minerals reveal, that a non-equilibrium state existed. Widmann-stätten figures prove, that diffusion took place on a long time scale. Then we see and also measure, that meteorites suffered several collisions (on the average about every 0.5 b.y.).Reheating can be seen in many components.

The ideal conditions for a K-Ar-age, where K^{40} decayed in situ and Ar^{40} is completely retained can therefore not be expected. In the interpretation of the ages this complex history has to be considered. The Ar/K ratios can be changed by large factors both by diffusion of K leaving radiogenic Ar behind and by diffusion losses of Ar^{40} mainly from inclusions. The differences between the true Ar/K and the observed Ar/K values are higher the later the diffusion took place, because the initial K^{40} abundance and the Ar^{40} production was 6.5 b.y. about 32 times higher. It is hoped that from such Ar/K variations more information on the history of meteorite evolution can be obtained. For the time being we have to conclude from these K-Ar-ages that the formation and cooling of the parent bodies of the iron meteorites started already about 6.3 b.y. ago. This age is higher than the age of the planetary system assumed so far. We probably have to admit that the history of the planetary system started earlier and that it is different for the earth and for the stony meteorites.

Literature

Abbey, S., and J. A. Maxwell: Determination of potassium in micas. Chem. in Can. 12, 37 (1960).

Adams, B. A., and E. L. Holmes: Absorptive properties of synthetic resins. J. Soc. Chem. Ind. (London) 54 T, 1 (1935).

Afanasyev, G. D., I. V. Borisevich, L. L. Shanin, and I. P. Sheina: Cases of unbalanced Ar-K ratios in biotites and their relevance to the establishment of an absolute geological time-scale: Intern. Geol. Rev. 7, 27 (1965).

— T. K. Kojina, and I. E. Starik: Results of age determination for standard muscovite, biotite and microline samples by the argon method. Rept. 21st Intern. Geol. Congr., Norden, Part 3, 28 (1960).

Ahrens, L. H., and R. D. Evans: The radioactive decay constants of K^{40} as determined from accumulation of Ca^{40} in ancient minerals. Phys. Rev. 74, 279 (1949).

Aldrich, L. T., and A. O. Nier: Argon 40 in potassium minerals. Phys. Rev. 74, 876 (1948).

— J. B. Doak, and G. L. Davis: The use of ion exchange columns in mineral analysis for age determination. Am. J. Sci. 251, 377 (1953).

— Measurement of radioactive ages of rocks. Science 123, 871 (1956).

— G. W. Wetherill, G. R. Tilton, and G. L. Davis: Half-life of Rb^{87}. Phys. Rev. 103, 1045 (1956a):

— G. L. Davis, G. R. Tilton, and G. W. Wetherill: Radioactive ages of minerals from the Brown Derby Mine and the Quartz Creek Granite near Gunnison, Colorado. J. Geophys. Res. 61, 215 (1956b).

— — — —, and P. M. Jeffery: Evaluation of Mineral Age Measurements. I., NAS-NRC publication 400, p. 147 (1956c).

—, G. W. Wetherill: Geochronology by radioactive decay. Ann. Rev. Nucl. Sci. 8, 257 (1958).

— — G. L. Davis, and G. R. Tilton: Radioactive ages of micas from granitic rocks by Rb-Sr and K-A methods. Trans. Am. Geophys. Union 39, 1124 (1958).

— — Rb-Sr and K-A ages of rocks in Ontario and Northern Minnesota. J. Geophys. Res. 65, 337 (1960).

Allen, E. T., and E. G. Zies: A chemical study of the fumaroles of the Katmai Region. Nat. Geog. Soc., Katmai Ser. No. 2, 75 (1923).

Allen, J. S.: The emission of secondary electrons from metals bombarded with protons. Phys. Rev. 55, 336 (1939a).

— The detection of single positive ions, electrons and photons by a secondary electron multiplier. Phys. Rev. 55, 966 (1939b).

ALLEN, P., M. H. DODSON, and D. C. REX: Potassium-argon dates and the origin of Wealden glauconites. Nature (Lond.) 202, No. 4932, 585 (1964).

ALLER, L. H.: The Abundance of the Elements. New York, London: Interscience Publishers 1961.

ALPERT, D.: New developments in the production and measurement of ultra high vacuum. J. Appl. Phys. 24, 860 (1953).

AMIRKHANOV, KH. I.: Rapid method to determine the absolute age of geologic formations by the radioactive decay of Potassium 40 to Argon 40 (in Russian). Bulletin of the 3rd Session of the Committee on the Absolute Age-Determination of the Geologic Formations 1954, Moscow 1955.

— I. G. GURVICH, and S. S. SARDAROV: Rapid mass-spectrometric method of determination of the absolute age of geologic formations according to the radioactive decay of K^{40} to Ar^{40} (in Russian). Izvest. Akad. Nauk USSR, Ser. Geol. 4, 80 (1955a).

— — L. L. SCHANIN, and S. S. SARDAROV: Mass-spectrometric method for measureing the quantity of radioargon in geological formation for a determination of their absolute age (in Russian). Zhur. Tekh. Fiz. 25, 558 (1955b).

—, and S. B. BRANDT: Determination of the absolute age of rocks according to the radioactive decay of K^{40} to Ar^{40} (in Russian). Akad. Nauk USSR., Machatschkala, Moscow 1956.

— K. S. MAGATAEV, and S. B. BRANDT: Determination of the absolute age of sedimentary minerals by radioactive methods. Akad. Nauk USSR Dokl. 117, No. 4, 675 (1957).

—, S. B. BRANDT, YE. N. BARNITZKII, V. S. GURVICH, and S. A. GASANOV: On the question of retention of radiogenic argon in glauconites. Akad. Nauk USSR Dokl. 118, No. 2, 328 (1958a).

— K. S. MAGATAEV, and G. I. TIMOFEYEVA: Results of the determination of the absolute age of sedimentary rocks of the oil-bearing provinces of Dagestan. Akad. Nauk USSR, Commission for the Determination of the Absolute Age of Geological Formations, 5th Sess., Trudy, 69 (1958b).

— E. N. BARTNITZKII, S. B. BRANDT, and G. V. VOITKEVICH: The migration of argon and helium in certain rocks and minerals. Doklady Akad. Sci. USSR 126, 160 (1959a).

— S. B. BRANDT, E. N. BARTNITZKII, and S. N. VORONOWSKII: The diffusion of radiogenic argon in sylvites. Doklady Akad. Nauk. USSR 126, 538 (1959b).

— — — The diffusion of radiogenic argon in feldspars. Doklady Akad. Nauk USSR 125, 210 (1959c).

—, and K. S. MAGATAEV: On the results of establishing the absolute age of sedimentary deposits in the oil-bearing region of Dagestan. Akad. Nauk USSR, Commission for the Determination of the Absolute Age of Geologic Formations, 6th sess., 48 (1960a).

— — V. S. IVANOV, and M. S. TRUZHNIKOV: On the method of mass-spectroscopic determination of radiogenic argon in rocks. Akad. Nauk USSR, Commission for the Determination of the Absolute Age of Geologic Formations, 6th sess., 287 (1960b).

— S. B. BRANDT, and E. N. BARTNITZKIJ: Radiogenic argon in minerals and rocks. Akad. Nauk. USSR, Machatschkala, Moscow, Leningrad (1960c).

ANDERS, E.: Meteorite Ages in "The moon, meteorites and comets" 402, Ed. MIDDLEHURST and KUIPER. Chicago: University of Chicago Press 1963.

— Origin, age, and composition of meteorites. Space Sci. Rev. 3, 583 (1964).

ARCHIBALD, E. H., W. G. WILCOX, and B. G. BUCKLEY: A study of the solubility of potassium chloroplatinate. J. Am. Chem. Soc. 30, 747 (1908).

ARMSTRONG, R. L., and E. HANSEN: Cordilleran infrastructure in the eastern Great Basin. Am. J. Sci. 112 (1965).
— Potassium argon dating of silicates using neutron activation for argon determination. Geol. Soc. Am. Bull. 70, 1563 (1959).
— Geochronology and geology of the eastern Great Basin in Nevada and Utah. Dissertation Yale University 1963.
ARNOLD, J. R.: Nuclear effects of cosmic rays in meteorites. Ann. Rev. Nucl. Sci. 11, 349 (1961).
— The origin of meteorites as small bodies II and III. Ap. J. 141, 1536 (1965).
ARROL, W. J., R. B. JACOBI, and F. A. PANETH: Meteorites and the age of the solar system. Nature (Lond.) 149, 235 (1942).
ARTEMOV, YU. M.: Absolute age of some rocks in the southern part of the Tarakskii massif (Yenesei Range). Geochemistry, no. 2, 191, Transl. from Geokhimiya Publ. Acad. Sci. USSR, no. 2, 174 (1963).
ASWATHANARAYANA, U.: Isotopic ages from the Eastern Ghats and Cuddapahs of India. J. Geophys. Res. 69, 3479 (1964).
AUER, S., H. J. BRAUN, and J. ZÄHRINGER: Die Edelgasisotope der Hauptbestandteile zweier Steinmeteorite. Z. Naturforsch. 20a, 156 (1965).
BAADSGAARD, H., S. S. GOLDICH, A. O. NIER, and J. H. HOFFMAN: The reproducibility of A^{40}/K^{40} determinations. Trans. Am. Geophys. Union 38, 539 (1957).
—, J. LIPSON, and R. E. FOLINSBEE: The leakage of radiogenic argon from sanidine. Geochim. Cosmochim. Acta 25, 147 (1961).
—, and M. H. DODSON: Potassium-argon ages of sedimentary and pyroclastic rocks. Quart. J. Geol. Soc. London, 120 S, 119 (1964).
— G. L. CUMMING, R. E. FOLINSBEE, and J. D. GODFREY: Limitations of radiometric dating. Roy. Soc. Can. Spec. Pub. 8, 20 (1964).
BACKENSTOSS, G., and K. GOEBEL: Die Zahl der Gamma-Quanten des Kaliums. Z. Naturforsch. 10a, 920 (1955).
BAILEY, S. W., P. M. HURLEY, H. W. FAIRBAIRN, and W. H. PINSON JR: K-Ar dating of sedimentary illite polytypes. Bull. Geol. Soc. Am. 73, 1167 (1962).
BAKER, C. A., and F. W. J. GARTON: A study of interferences in emission and absorption flame photometry. Report AERE-R- 3491, Atomic Energy Research Establishment, England (1961).
BAKER, G.: Form and sculpture of tektites, in "Tektites", 141. Ed. J. O'KEEFE, Chicago: University of Chicago Press 1963.
BALESTRINI, S. J.: Mass-spectronometric study of deuteron induced reactions in iodine. Phys. Rev. 95, 1502 (1954).
BARNES, V. E.: Tektite strewn fields, in "Tektites", 25. Ed. J. O'KEEFE, Chicago: University of Chicago Press 1963.
BARRINGTON, A. E.: High vacuum engineering. Englewood Chiffs, N. J.: Prentice Hall Inc. 1963.
BASSETT, W. A., P. F. KERR, O. A. SCHAEFFER, and R. W. STOENNER: Potassium-argon ages of volcanics near Grants, New Mexico. Bull. Geol. Soc. Am. 74, 213 (1963).
BATE, G. L., J. R. HIUZENGA, and H. A. POTRAZ: Thorium in stone meteorites by neutron activation analysis. Geochim. Cosmochim. Acta 16, 88 (1959).
BAUER, C. A.: Production of He in meteorites by cosmic radiation. Phys. Rev. 72, 345 (1947).
BAUMANN, E. W., and W. J. JR. ARGERSINGER: Influence of electrolyte uptake on the equilibrium constant for the sodium-hydrogen exchange on Dowex 50. J. Am. Chem. Soc. 78, 1130 (1956).

BEALL, G. H.: Age of authigenic biotite in the Utica shale. Am. Min. **47**, 971 (1962).

BEARD, G. B., and W. H. KELLY: Die Halbwertszeit von Rb^{87}. Nucl. Phys. **28**, 570 (1961).

BECK, A. H. (Editor): Handbook of vacuum physics. Oxford: Pergamon Press 1964.

BEGEMANN, F., J. GEISS, and D. C. HESS: Radiation age of meteorite from cosmic-ray produced H^3 and He^3. Phys. Rev. **107**, 540 (1957).

— P. EBERHARDT, D. C. HESS: ^3He-^3H-Strahlungsalter eines Steinmeteoriten. Z. Naturforsch. **14a**, 500 (1959).

BEISER, E., and W. H. PINSON: Rb-Sr age of the Murray carbonaceous chondrite. Trans. Am. Geophys. Union **45**, 91 (1964).

BELLAIR, P., et L. DELBOS: Age absolu de la dernière granitisation en Terre Adelie. C. R. Acad. Sci. (Paris) **254**, 1465 (1962).

BEVERIDGE, A. J., and R. E. FOLINSBEE: Dating Cordilleran orogenies. J. Alberta Soc. Petrol. Geol. **4**, 236 (1956).

BEYER, H. O.: The relation of Tektites to Archaeology. Phil. Nat. Res. Council, Univ., Phil. Publ. Quezon City: Dillman 1955.

BIERI, R. H., and W. RUTSCH: The Diffusion of noble gases in the chondrites "Bruderheim" and "Holbrook". (1965) (to be published).

BINNS, R. A., and J. A. MILLER: Potassium-argon determinations on some rocks from the Broken Hill region of New South Wales. Nature (Lond.) **199**, 274 (1963).

BIRCH, F.: Geophysics and the radioactivity of potassium. Phys. Rev. **72**, 1128 (1947).

BIRKENFELD, H., G. HAASE, and H. ZAHN: Massenspektrometrische Isotopen-analyse. Berlin: VEB Deutscher Verlag der Wissenschaft 1962.

BIRKS, L. S.: X-Ray Spectrochemical Analysis, Chemical Analysis. Vol. XI. New York: Interscience Publishers Inc. 1964.

BLEULER, E., and M. GABRIEL: Zerfall des K^{40}. Helv. Phys. Acta **20**, 67 (1947).

BORST, L. B., and J. J. FLOYD: The radioactive decay of K^{40}. Phys. Rev. **74**, 989 (1948).

BOYD, G. E.: Method of activation analysis. Analyt. Chem. **21**, 335 (1949).

BRANDT, S. B., and S. N. VORONOVSKY: Dehydration and diffusion of radiogenic argon from micas. Akad. Nauk USSR **11**, 78 (1964a).

— — Dependence of the energy of activation of radiogenic gases in minerals on temperature. Geokhimiya **4**, 375 (1964b).

BREZINA, A.: The arrangement of collections of meteorites. Proc. Am. Phil. Soc. **43**, 211 (1904).

BROWN, P. E., J. A. MILLER, and N. J. SOPER: Age of the principal intrusion of the Lake District. Proc. Yorkshire Geol. Soc. **34**, 331 (1964).

BRUNNÉE, C., and H. VOSHAGE: Massenspektrometrie. München: Thiemig Verlag 1964.

BURCH, P. R. A.: Specific gamma-activity, the branching ratio and half-life of potassium 40. Nature (Lond.) **172**, 361 (1953).

BURGESS, L. L., and O. KAMM: A study of cobaltinitrites and their application to analytical chemistry. J. Am. Chem. Soc. **34**, 652 (1912).

BURNETT, D. S., A. W. FOWLER, and F. HOYLE: Nucleosynthesis in the early history of the solar system. Geochim. Cosmochim. Acta **29**, 1209 (1965).

— H. J. LIPPOLT, and G. J. WASSERBURG: The relative isotopic abundance of K^{40} in terrestrial and meteoritic samples. J. Geophys. Res. **71**, 1249 (1966).

BURRIEL-MARTI, F., and J. RAMIREZ-MUNOZ: Flame Photometry. New York: Elsevier 1957.

BURWASH, R. A., H. BAADSGAARD, F. A. CAMPBELL, G. L. CUMMING, and R. E. FOLINSBEE: Potassium-argon dates of diabase dyke systems, District of Mackenzie, N. W. T. (Canada). Can. Mining Met. Bull. **56**, 706 (1963).

BUTCHER, N. E.: Age of the orogeny and granites in southwest England. Nature (Lond.) **190**, 253 (1961).

BYSTROEMASKLUND, A. M., H. BAADSGAARD, and R. E. FOLINSBEE: K-Ar age of biotite, sanidine and illite from Middle Ordovician bentonites at Kinnekulle, Sweden. Geol. Fören. i Stockholm Förh. **83**, 92 (1961).

CAMERON, R. L., S. S. GOLDICH, and J. H. HOFFMAN: Radioactivity age of rocks from the Windmill Islands, Budd Coast, Antartica. Stockholm Contr. Geol. **6**, 1 (1960).

CAMERON, A. G. W.: The formation of the sun and planets. Icarus **1**, 13 (1962).

CARR, D. R., and J. L. KULP: Use of A^{37} to determine argon behavior in vacuum systems. Rev. Sci. Instr. **26**, 379 (1955).

— — Potassium-argon-method of geochronometry. Bull. Geol. Soc. Am. **68**, 763 (1957).

CARSLAW, H. S., and J. C. JAEGER: Conduction of heat in solids (2nd Ed.). Oxford: Clarendon Press 1959.

CHAPMAN, D. R., and H. K. LARSON: The lunar origin of tektites. NASA, TND-1556 (1963).

CHERDYNTSEV, V. V., and YE. M. KOLESNIKOV: The argon of ancient sedimentary rocks. Geokhimiya **9**, 8 49 (1964).

CLARKE, R. S. JR., and E. P. HENDERSON: Georgia tektites and related glasses. Georgia Mineral Newsletter **14**, 90 (1961).

COHEN, E.: Metoritenkunde I—III. Stuttgart: Schweizbart E. 1894, 1903, 1905.

COHN, W. E., and H. W. KOHN: Ion-exchange separation of the alkali metals. J. Am. Chem. Soc. **70**, 1986 (1948).

COMPSTON, W., and M. J. VERNON: Personal communication to Cooper (1963).

COOK, L. G.: Untersuchungen über Chrom- und Eisenhydroxyde und ihre Bedeutung für die Emaniermethode. Z. phys. Chem. **42**, 221 (1939).

COOPER, J. A.: The flame photometric determination of potassium in geological materials used for potassium argon dating. Geochim. Cosmochim. Acta **27**, 525 (1963).

— J. R. RICHARDS, and A. W. WEBB: Some potassium-argon ages in New England, New South Wales. J. Geol. Soc. Austr. **10**, 313 (1963).

CORNISH, F. W.: The practical application of chromatographic theory to analytical and preparative separations by ion-exchange elution. Analyst **83**, 634 (1958).

— In "Methods in Geochemistry" edited by SMALES and WAGER, New York, London: Interscience Publishers 1960.

CURRAN, S. C.: The determination of geological age by means of radioactivity. Quart. Rev. (Lond.) **7**, 1 (1953).

CURTIS, G. H.: Mode of origin of pyroclastic debris in the Mehrten formation of the Sierra Nevada. Univ. Calif. Publ. Bull. Dep. Geol. Sci. **29**, 453 (1954).

—, and J. H. REYNOLDS: Notes on the potassium-argon dating of sedimentary rocks. Bull. Geol. Soc. Am. **69**, 151 (1958).

— D. E. SAVAGE, and J. F. EVERNDEN: Critical points in the Cenozoic. N. Y. Acad. Sci. Ann. **91**, 342 (1961).

DALRYMPLE, G. B.: Argon retention in a granitic xenolith from a Pleistocene basalt, Sierra Nevada, California. Nature (Lond.) **201**, 282 (1964).

—, and K. HIROOKA: Variation of potassium, argon, and calculated age in a late Tertiary basalt flow. Trans. Am. Geophys. Union **46**, 178 (1965).

DAMON, P. E., and J. L. KULP: Determination of radiogenic helium in zircon by stable isotope dilution technique. Trans. Am. Geophys. Union **38**, 945 (1957a).

— — Argon in mica and the age of Beryl Mt., N. H., pegmatite. Am. J. Sci. **255**, 697 (1957b).

DAMON, P. E., and J. L. KULP: Excess helium and argon in beryl and other minerals. Am. Mineralogist 43, 433 (1958).
— Correlation and chronology of ore deposits and volcanic rocks. Ann. Progr. Rep. 4, 11 (1962) (University of Arizona).
DAVIDSON, C. F.: The geological time scale. Nature (Lond.) 184, 1473 (1959).
DAVIS, G. L., G. R. TILTON, and G. W. WETHERILL: Mineral ages from the Appalachian Province in North Carolina and Tennessee. J. Geophys. Res. 67, 1987 (1962).
DEAN, J. A.: Flame Photometry. New York: McGraw-Hill 1960.
DELANEY, C. F. G.: Beta-activity of potassium. Phys. Rev. 82, 158 (1951).
DEN TEX, E.: A commentary on the correlation of metamorphism and deformation in space and time. Geol. Mijnbouw 42, 170 (1963).
DEUTSCH, S., and L. CHAURIS: Mesure de l'age absolu du granite du Huelgoat Finistère. C. R. Acad. Sci. (Paris) 250, 1298 (1960).
DIELS, K., and R. JAECKEL: Leybold Vakuum Taschenbuch. Berlin-Göttingen-Heidelberg: Springer-Verlag 1962.
DIENES, G. J., and G. H. VINEYARD: Radiation effects in solids. New York, London: Interscience Publishers 1957.
DODSON, M. H., J. A. MILLER, and D. YORK: Potassium-argon ages of the Dartmoor and Shap granites using the total volume and isotopic dilution techniques of argon measurement. Nature (Lond.) 190, 800 (1961).
— Further argon age determinations on slates from southwest England. Proc. Usher Soc. 1, 70 (1963).
— D. C. REX, R. CASEY, and P. ALLEN: Glauconite dates from the Upper Jurassic and Lower Cretaceous. Quart. J. Geol. Soc. London 1208, 145 (1964).
DOE, B. R.: Relationship of lead isotopes among granites, pegmatites, and sulfide ores near Balmat, New York. J. Geophys. Res. 67, 2895 (1962).
DUBANSKY, A.: Determination of the absolute age of rocks by the potassium-argon method. Trans. Cz. Acad. Sci. ser. Math. Natur. 73, 77 (1963).
DUBAR, G. P.: Combined use of age data and mineral analysis in paleographic problems, with reference to the basin of the Lena. Dokl. Acad. Sci. USSR, Earth Sci. Sect., 143, 13—15. Transl. from Dokl. Akad. Nauk USSR 143, 175 (1962).
DUCKWORTH, H. E.: Mass spectroscopy. Cambridge: University Press 1958.
DUSHMANN, S.: Scientific foundations of vacuum technique (2nd edn.). London, New York: Wiley 1962.
DZELEPOW, B., M. KOPJOVA, and E. VOROBJOV: β-ray spectrum of K⁴⁰. Phys. Rev. 69, 538 (1946).
EBERHARDT, P., and D. C. HESS: Helium in stone meteorites. Astrophys. J. 131, 38 (1960).
— J. GEISS, F. G. HOUTERMANS, and P. SIGNER: Age determinations on lead ores. Geol. Rundsch. 52, 836 (1962).
— — Meteorite classes and radiation ages in "Isotopic and cosmic chemistry". Amsterdam: North Holland Publishing Co. 1964.
— O. EUGSTER, and J. GEISS: Radiation ages of aubrites. J. Geophys. Res. 70, 4427 (1965).
— O. EUGSTER, J. GEISS, and K. MARTI: Rare gas measurements in 30 stone meteorites. Z. Naturforsch. 21a, 414 (1966).
EDWARDS, G.: Sodium and potassium in meteorites. Geochim. Cosmochim. Acta 8, 285 (1955).
—, and H. C. UREY: Determination of alkali metals in meteorites by a distillation process. Geochim. Cosmochim. Acta 7, 154 (1955).
EGELKRAUT, K., u. H. LEUTZ: Halbwertzeit des K⁴⁰. Phys. Verhandl. 11, 67 (1960).

EKEDAHL, E., E. HÖGFELDT, and L. G. SILLEN: Ion-exchange equilibria. Acta Chem. Scand. 4, 556 and 4, 828 (1950).

ELGORESY, A.: Mineralbestand und Strukturen der Graphit- und Sulfideinschlüsse in Eisenmeteoriten. Geochim. Cosmochim. Acta 29, 1131 (1965).

ELLIOTT, H. M. (Editor): Advances in mass spectrometry II. London: Pergamon Press 1963.

EMSLIE, R. F.: Potassium-argon age of the Michikamau anorthositic intrusion, Labrador. Nature (Lond.) 202, 172 (1964).

ENDT, P. M., and J. C. KLUYVER: Energy levels of light nuclei ($Z = 11$ to $Z = 20$) Rev. Mod. Phys. 26, 96 (1954).

ERICKSON, G. P., and J. L. KULP: Potassium-argon measurements on the Palisades sill, New Jersey. Geol. Soc. Am. Bull. 72, 649 (1961).

— — K-Ar dating of basalts (abs.). Geol Soc. America Spec. Paper 76, 55 (1964).

EVANS, C. R., J. STEINER, and J. L. WEINER: Age-dating studies on Precambrian rocks at Jasper, Alberta. Can. Mining Met. Bull. 57, 33 (1964).

EVERNDEN, J. F., G. H. CURTIS, and J. LIPSON: Potassium-Argon dating on igneous rocks. Am. Assoc. Petroleum Geol. Bull. 41, 2120 (1957).

EVERNDEN, J. F., G. H. CURTIS, R. W. KISTLER, and J. OBRADOVICH: Argon diffusion in glauconite, microcline, sanidine, leucite and phlogopite. Am. J. Sci. 258 B, 583 (1960).

— — J. OBRADOVICH, and R. KISTLER: On the evaluation of glauconite and illite for dating sedimentary rocks by the potassium-argon method. Geochim. Cosmochim. Acta 23, 78 (1961).

—, and J. R. RICHARDS: Potassium-argon ages at Broken Hill, Australia. Nature (Lond.) 192, 446 (1961).

— — Potassium argon ages in Eastern Australia. J. Geol. Soc. Aust. 9, 1 (1962).

— D. E. SAVAGE, G. H. CURTIS, and G. T. JAMES: Potassium-argon dates and the Cenozoic mammalian chronology of North America. Am. J. Sci. 262, 145 (1964).

—, and G. H. CURTIS: The potassium-argon dating of Late Cenozoic Rocks in East Africa and Italy. Current Anthropology 6, 343 (1965).

EWALD, H., u. H. HINTENBERGER: Methoden und Anwendungen der Massen-spektroskopie. Weinheim: Verlag Chemie 1953.

EWING, G. W.: Instrumental Methods of Chemical Analysis. New York: McGraw-Hill Book Company 1960.

FAHRIG, W. F., and R. K. WANLESS: Age and significance of diabase dyke swarms of the Canadian Shield. Nature (Lond.) 200, 934 (1963).

FAIRBAIRN, H. W., P. M. HURLEY, W. H. JR. PINSON, and R. F. CORMIER: Age of the granitic rocks of Nova Scotia. Bull. Geol. Soc. Am. 71, 399 (1960a).

— W. H. PINSON, P. M. HURLEY, and R. F. CORMIER: A comparison of the ages of coexisting biotite and muscovite in some Paleozoic granitic rocks. Geochim. Cosmochim. Acta 19, 7 (1960b).

FANALE, F. P., and J. L. KULP: The helium method and the age of the Cornwall Pennsylvania Magnetic Ore. Econ. Geol. 57, 735 (1962).

FARRAR, R. L., and G. H. CADY: Comparison of age with the relative abundance of potassium and argon in rocks. J. Am. Chem. Soc. 71, 742 (1949).

— R. M. MACINTYRE, D. YORK, and W. J. KENYON: A simple mass spectrometer for analysis of argon at ultra-high vacuum. Nature (Lond.) 204, 531 (1964).

FAUL, H.: Age of the Fen carbonatite (Norway) and its relation to the intrusives of the Oslo region. Geochim. Cosmochim. Acta 17, 153 (1959).

— Geologic time scale. Bull. Geol. Soc. Am. 71, 637 (1960).

—, and E. JAEGER: Ages of some granitic rocks in Vosges, the Schwarzwald, and the Massif Central. J. Geophys. Res. 68, 3293 (1963).

FAUST, H., T. W. STERN, H. H. THOMAS, and P. L. D. ELMORE: Ages of intrusion and metamorphism in the northern Appalachians. Am. J. Sci. 261, 1 (1963).

FAUST, W. R.: Specific activity of potassium. Phys. Rev. 78, 624 (1950).

FECHTIG, H., W. GENTNER and J. ZÄHRINGER: Argonbestimmungen an Kaliummineralien — VII. Diffusionsverluste von Argon in Mineralien und ihre Auswirkung auf die Kalium-Argon-Altersbestimmung. Geochim. Cosmochim. Acta 19, 70 (1960).

— — and S. KALBITZER: Argonbestimmungen an Kaliummineralien — IX. Messungen zu den verschiedenen Arten der Argondiffusion. Geochim. Cosmochim. Acta 25, 297 (1961).

— — and P. LÄMMERZAHL: Argonbestimmungen an Kaliummineralien — XII Edelgasdiffusionsmessungen an Stein- und Eisenmeteoriten. Geochim. Cosmochim. Acta 27, 1149 (1963).

FELIX, F., T. LAGERWALL, P. SCHMELING, and K. E. ZIMEN: Rare Gas Release from Solids with Special Attention to Nuclear Fuel Materials. EUR 1506.e EURAEC Report Nr. 906 (1964).

FENNER, C.: Australites; Part III. A contribution to the problem of the origin of tektites. Roy. Soc. South Australia, Trans. 62, 192 (1938).

FIREMAN, E. L., and D. SCHWARZER: Measurement of Li⁶, He³, and H³ in meteorites and its relation to cosmic radiation. Geochim. Cosmochim. Acta 11, 252 (1957).

FISHER, N. H.: Age determinations of northern Australian granites. Rept. 21st Intern, Geol. Congr., Norden, Part 9, 179 (1960).

FISHER, D. E.: Ages of the Sikhote Alin iron meteorite. Science 139, 752 (1963).

— Anomalous Ar⁴⁰ content in iron meteorites. J. Geophys. Res. 70, 2445 (1965).

FISHER, R. V.: Classifikation of volcanic breccias. Bull. Geol. Soc. Am. 71, 923 (1960).

FISHMAN, M. V.: New data on the age of granitoid plutons of the northern (circumpolar) Urals. Dokl. Acad. Sci. USSR, Earth Sci. Sect., 145, 53. Transl. from Dokl. Akad. Nauk USSR 145, 400 (1962).

FLEISCHER, M., and R. E. STEVENS: Summary of new data on rock samples G-1 and W-1. Geochim. Cosmochim. Acta 26, 525 (1962).

FLEISCHER, R. L., and P. B. PRICE: Techniques for geological dating of minerals by chemical etching of fission fragment tracks. Geochim. Cosmochim. Acta 28, 1705 (1964).

FLÜGGE, S., and K. G. ZIEMENS: Die Bestimmung von Korngrößen und von Diffusionskonstanten aus dem Emaniervermögen. Z. phys. Chem. 42, 179 (1939).

FLYNN, K. F., and L. E. GLENDENIN: The half-life and beta spectrum of Rb⁸⁷. Phys. Rev. 116, 744 (1959).

FOLINSBEE, R. E., J. LIPSON, and J. H. REYNOLDS: Potassium-argon dating. Geochim. Cosmochim. Acta 10, 60 (1956).

— H. BAADSGAARD, and G. L. CUMMING: Dating of volcanic ash beds (bentonites) by the K-Ar method. In Nuclear Geophysics, Nat. Acad. Sci. Nat. Res. Council Nucl. Sci. Ser. Rep. 38, 70 (1962).

— —, and J. LIPSON: Potassium-argon time scale. Rept. 21st. Intern. Geol. Congr., Norden, Part 3, 7 (1960).

— — — Potassium-argon dates of Upper Cretaceous ash falls. Alberta, Canada. N. Y. Acad. Sci. Ann. 91, 352 (1961).

FOWLER, W. A., J. L. GREENSTEIN, and F. HOYLE: Nucleosynthesis during the early history of the solar system. Geophys. J. Roy. Astr. Soc. 6, 148 (1962).

FRECHEN, J., u. H. J. LIPPOLT: Kalium-Argon-Daten zum Alter des Laacher Vulkanismus, der Rheinterrassen und der Eiszeiten. Eiszeitalter u. Gegenwart 16, 5 (1965).

FREDERIKSSON, K., P. S. DE CARLI, and A. AARAMAE: Shock-induced veins in chondrites. Space Research III, 974. Amsterdam: Priester, North-Holland Publishing Co. 1963.

FRISOV, L. V.: Age of metamorphic rocks in the northeast of the USSR and signs of repeated metamorphism. Dokl. Acad. Sci. USSR, Earth Sci. Sect. 143, 32. Transl. from Dokl. Akad. Nauk USSR 143, 204 (1962).
— Late Paleozoic igneous activity in northeast USSR. Dokl. Acad. Sci. USSR, Earth Sci. Sect. 142, 68 (1964). Transl. from Dokl. Akad. Nauk USSR 142, 913 (1962b).
FURMAN, N. H. (Editor): Scott's Standard Methods of Chemical Analysis 6th edition, 1, The Elements D. Princeton (New Jersey), New York: Van Nostrand Company 1962.
GABRIELSE, H., and J. E. REESOR: Geochronology of plutonic rocks in two areas of the Canadian Cordillera. Roy. Soc. Spec. Canada, Pub. 8, 96 (1964).
GARETZKY, R. G., YE. M. KOLESNIKOV, V. I. MURAV'YEV, and A. YE. SHLEZINGER: On the possibility of determining the absolute age of folding according to authigenic minerals in sedimentary rocks (exemplified by the folded basement of the southern Aral area). Akad. Nauk. USSR Doklady 154, 4 (1964).
GAST, P. W., J. L. KULP, and L. E. LONG: Absolute age of early Precambrian rocks in the Bighorn Basin of Wyoming and Montana, and Southeastern Manitoba. Trans. Am. Geophys. Union 39, 322 (1958).
— Limitations on the composition of the upper mantle. J. Geophys. Res. 65, 1287 (1960).
— Strontium and rubidium in stone meteorites. Problems related to interplanetary matter. NAS-NRC publ. 845, 85 (1961).
— The isotopic composition of strontium and the age of stone meteorites. — I. Geochim. Cosmochim. Acta 26, 927 (1962).
— Rb, Sr, Ba, and isotopic composition of Sr in some stone meteorites. Trans. Am. Geophys. Union 44, 87 (1963).
GEILMANN, W., u. W. GEBAUHR: Zur Fällung der Alkalimetalle als Tetraphenyl-borverbindungen. Z. anal. Chem. 139, 161 (1953).
GEISS, J., and D. C. HESS: Argon-potassium ages and the isotopic composition of argon from meteorites. Astrophys. J. 127, 224 (1958).
— H. OESCHGER, and P. SIGNER: Radiation ages of chondrites. Z. Naturforsch. 15a, 1016 (1960).
— —, and U. SCHWARZ: The history of cosmic radiation as revealed by isotopic changes in the meteorites and on earth. Proc. Int. School of Physics "Enrico Fermi" XIX, Solar Particles and Space Research, May 1961, Acad. Press New York (1963) p. 247, and also Space Sci. Rev. (1962) 1, 197 (1961).
GENTNER, W., R. PRÄG, and F. SMITS: Argonbestimmungen an Kaliummineralien, II. Das Alter eines Kalilagers im unteren Oligozän. Geochim. Cosmochim. Acta 4, 11 (1953a).
— — — Altersbestimmungen nach der Kalium-Argonmethode unter Berück-sichtigung der Diffusion des Argons. Z. Naturforsch. 8a, 216 (1953b).
— F. JENSEN, and K. R. MEHNERT: Zur geologischen Altersbestimmung von Kali-feldspat nach der Kalium-Argon-Methode. Z. Naturforsch. 9a, 176 (1954a).
— K. GOEBEL, and R. PRÄG: Argonbestimmungen an Kaliummineralien, III: Ver-gleichende Messungen nach der Kalium-Argon- und -Uran-Helium-Methode. Geochim. Cosmochim. Acta 5, 124 (1954b).
—, u. E. A. TRENDELENBURG: Experimentelle Untersuchungen über die Diffusion von Helium in Steinsalzen und Sylvinen. Geochim. Cosmochim. Acta 6, 261 (1954).
—, and W. KLEY: On the geological determination of age by the potassium-argon method. Z. Naturforsch. 10a, 832 (1955).
—, u. J. ZÄHRINGER: Argon und Heliumbestimmungen in Eisenmeteoriten. Z. Naturforsch. 10a, 498 (1955).

GENTNER, W., and W. KLEY: Argonbestimmungen an Kaliummineralien — IV. Die Frage der Argonverluste in Kalifeldspäten und Glimmermineralien. Geochim. Cosmochim. Acta 12, 323 (1957).

—, and J. ZÄHRINGER: Argon und Helium als Kernreaktionsprodukte in Meteoriten. Geochim Cosmochim. Acta 11, 60 (1957).

—, u. W. KLEY: Altersbestimmungen nach der Kalium-Argon-Methode an Mineralien und Gesteinen des Schwarzwaldes. Geochim. Cosmochim Acta 14, 98 (1958).

—, u. J. ZÄHRINGER: K-Ar Alter einige Tektite. Z. Naturforsch. 14a, 686 (1959).

— — Das Kalium-Argon-Alter von Tektiten. Z. Naturforsch. 15a, 93 (1960).

— H. J. LIPPOLT, and O. A. SCHAEFFER: Das Kalium-Argon-Alter einer Glasprobe vom Nördlinger Ries. Z. Naturforsch. 16a, 1240 (1961).

— — — Argonbestimmungen an Kaliummineralien. XI. Die Kalium-Argon-Alter der Gläser des Nördlinger Rieses und der böhmisch-mährischen Tektite. Geochim. Cosmochim. Acta 27, 191 (1963).

— — The Potassium-Argon Dating of Upper Tertiary and Pleistocene Deposits, in "Science in Archaeology", p. 72. London: Thames and Hudson 1963.

— —, and O. MÜLLER: Kalium-Argon-Alter des Bosumtwi-Kraters in Ghana und die chemische Beschaffenheit seiner Gläser. Z. Naturforsch. 19a, 150 (1964).

— Das Rätseln um die Herkunft der Tektite, in „Jahrb. der Max-Planck-Gesellschaft", 90 (1964).

GERLING, E. K.: Diffusionswärme des Heliums als Kriterium für die Brauchbarkeit von Mineralien für Altersbestimmungen nach der Heliummethode. C. R. Akad. Sci. USSR 24, 570 (1939).

—, and N. E. TITOV: Über den K-Zerfall des Kaliums (in Russian). Izv. Akad. Nauk USSR 2, 128 (1949).

— N. E. TITOV, and G. M. ERMOLIN: Determination of the K-capture constant of K⁴⁰. Dokl. Akad. Nauk, USSR 68, 553 (1949).

—, and T. G. PAVLOVA: Determination of the geological age of two stony meteorites by the argon method. Dokl. Akad. Nauk USSR 77, 85 (1951).

—, and K. G. RIK: On the forms of occurence of argon in meteorites. Dokl. Akad. Nauk USSR 101, 433 (1955).

—, and L. K. LEVSKII: On the origin of the rare gases in stony meteorites. Dokl. Nauk USSR 110, 750 (1956).

—, and I. M. MOROZOWA: Determination of the activation energy of argon liberated from micas. Geochemistry 4, 359 (1957).

— Effect of metamorphism on geologic age as determined by the lead method. Geochemistry, 363 (1958).

— — The kinetics of argon liberation from microline-perthite. Geochemistry 7, 775 (1958).

— Der heutige Stand der Argonmethode zur Altersbestimmung und ihre Anwendung in der Geologie (in Russian). Akad. Nauk USSR Moscow (1961).

— I. M. MOROZOVA, and V. V. KURBATOV: The retentivity of radiogenic argon in ground micas. N. Y. Acad. Sci. Annals 91, 227 (1961a).

— — — On the preservation state of radiogenic argon in ground potassium-containing minerals. Akad. Nauk USSR 1, 39 (1961b).

— — Determination of the activation energy for the release of argon and helium from minerals. Geochemistry 12, 1255 (1962).

—, and G. V. OVCHINNIKOVA: Causes of low age values determined on micas by the Rb-Sr method. Geochemistry, no. 9, 865, Transl. from Geokhimiya Publ. Acad. Sci. USSR, no. 9, 755 (1962).

GERLING, E. K., YU. A. SHUKOLYUKOV, T. V. KOL'TSOVA, I. I. MATVEYEVA, and
S. Z. YAKOVLEVA: Dating of mafic rocks by K/Ar method. Geochemistry, no.
11, 1055. Transl. from Geokhimiya Publ. Acad. Sci. USSR, no. 11, 931 (1962a).
— YU. A. SHUKOLYUKOV, and I. I. MATVEYVA: Determination of ages of beryl and
other minerals containing inclusions by the Rb-Sr method. Geochemistry, no. 1,
74. Transl. from Geokhimija Publ. Acad. Sci. USSR, no. 1, 67 (1962b).
— L. K. LEVSKII, and I. M. MOROZOVA: On the diffusion of radiogenic argon from
minerals. Geochemistry 6, 551 (1963).
—, and YU. A. SHUKOLYUKOV: Calculation of the time of differentiation of primary
magma using K-Ar data. Geochemistry, no. 4, 365. Transl. from Geokhimija
Publ. Acad. Sci. USSR no. 4, 347 (1963).
GIBBONS, D., B. H. LOVERIDGE, and R. J. MILLETT: Radioactivation analysis —
A bibliography (1957) A.E.R.E. J/R 2208, U.K.A.E.A. (1957).
GIFFIN, C. E., and J. L. KULP: Potassium-argon ages in the Precambrian basement
of Colorado. Bull. Geol. Soc. Am. 71, 219 (1960).
GILBERT, C. M.: Welded tuff in eastern California. Bull. Geol. Soc. Am. 49, 1829
(1938).
GILETTI, B. J., S. MOORBATH, and R. STJ. LAMBERT: A geochronological study of
the metamorphic complexes of the Scottish Highlands. Quart. J. Geol. Soc. 117,
233 (1961).
GLAESSNER, M. F.: The dating of the base of the Cambrian. J. Geol. Soc. India 4,
1 (1963).
GLEDITSCH, E., and T. GRAF: On the gamma-rays of K[40]. Phys. Rev. 72, 640 (1947).
GOEBEL, K., u. P. SCHMIDLIN: Tritium-Messungen an Steinmeteoriten. Z. Natur-
forsch. 15a, 79 (1960).
GOLDICH, S. S., H. BAADSGAARD, and A. O. NIER: Investigations in Ar[40]/K[40] dating.
Trans. Am. Geophys. Union 38, 547 (1957).
— — G. EDWARDS, and C. E. WEAVER: Investigations in radioactivity dating of
sediments. Bull. Am. Ass. Petrol. Geol. 43, 654 (1959).
— A. O. NIER, H. BAADSGAARD, H. HOFFMAN, and W. KRUEGER: The Precambrian
geology and geochronology of Minnesota. Minnesota Geol. Survey Bull. 41, 193
(1961).
GOLDSCHMIDT, V. M.: Geochemische Verteilungsgesetze der Elemente. XI. Die
Mengenverhältnisse der Elemente und der Atom-Arten. Norske Videnskaps —
Akad. Skrifter, Math.-Naturw. Klasse 4, 148 (1937).
GOLES, G. G., R. A. FISH, and E. ANDERS: The record in the meteorites — I. The
former environment of stone meteorites as deduced from K[40]-Ar[40] ages.
Geochim. Cosmochim. Acta 19, 177 (1960).
—, and E. ANDERS: Abundances of iodine, tellurium and uranium in meteorites.
Geochim. Cosmochim. Acta 26, 723 (1962).
GOOD, M. L., and R. D. EVANS: Beta-ray spectrum of K[40]. Phys. Rev. 83, 1054
(1941).
GOODMAN, C., and R. D. EVANS: Age measurements by radioactivity. Bull. Geol.
Soc. Am. 52, 491 (1941).
GRAF, T.: Significance of the radioactivity of potassium in geophysics II. Phys.
Rev. 74, 831 (1948).
— Construction and performance of a multiple gamma-ray counter of high effi-
ciency. Rev. Sci. Instr. 21, 285 (1950).
GREGOR, H. P.: Gibbs-Donnan equilibrium in ion exchange resin systems. J. Am.
Chem. Soc. 73, 642 (1951).
GULDNER, W. G., and A. L. BEACH: Vacuum fusion furnace for analysis of gases in
metals. Analyt. Chemistry 22, 366 (1950).

GUTHRIE, A.: Vacuum Technology. New York: Wiley 1963.

HADLEY, J. B.: Correlation of isotopic ages, crustal heating and sedimentation in the Appalachian region: V.P.I. Dept. Geol. Sciences Mem. I — Tectonics of the Southern Appalachians, p. 33 (1964).

HAHN-WEINHEIMER, P., and H. ACKERMANN: Quantitative röntgenspektral-analytische Bestimmung von Kalium, Rubidium, Strontium, Barium, Titan, Zirkonium und Phosphor. Z. analyt. Chem. 194, 81 (1963).

HAMAGUCHI, H., G. W. REED, and A. TURKEVICH: Uranium and barium in stone meteorites. Geochim. Cosmochim. Acta 12, 337 (1957).

HARPER, C. T.: Potassium-argon ages of slates and their geological significance. Nature 203, 468 (1964).

HART, S. R.: The use of hornblendes and pyroxenes for K-Ar dating. J. Geophys. Res. 66, 2995 (1961).

—, and R. T. DODD: Excess radiogenic argon in pyroxenes. J. Geophys. Res. 61, 2998 (1962).

— Excess argon in pyroxenes. U.S.N.A.C. — NRC Publ. 1075, Nuclear Sci. Ser. Rept. no. 38, 68 (1963).

— The petrology and isotopic-mineral age relations of a contact zone in the Front Range, Colorado. J. Geol. 72, 493 (1964).

HARTECK, P., u. H. E. SUESS: Der Argongehalt kalihaltiger Minerale und die Frage des dualen Zerfalls von K^{40}. Naturwissenschaften 34, 214 (1947).

HARTMANN, W., and F. BERNHARD: Photovervielfacher. Berlin: Akademie Verlag 1957.

HAVRE, G. N.: The flame photometric determination of sodium, potassium and calcium in plant extracts with special reference to interference effects. Analyt. Chim. Acta 25, 557 (1961).

HAY, R. L.: Stratigraphy and zeolitic diagenesis of the John Day formation of Oregon. Univ. Calif. Publ. Geol. Sci. 42, 199 (1963).

HAYDEN, R. J., and M. G. INGHRAM: Further results on the double beta decay of Te^{130}. U.S. Nat. Bur. Stand. Circ. 522, 189 (1953).

—, and J. P. WEHRENBERG: A^{40}-K^{40} dating of igenous and metamorphic rocks in western Montana. J. Geol. 68, 94 (1960).

HERR, W.: Neutron activation applied to geochemistry. Proc. of the radioactivation analysis symposium, p. 35 Vienna 1959. London: Butterworths 1960.

— W. HOFFMEISTER, B. HIRT, J. GEISS, and F. G. HOUTERMANS: Versuch zur Datierung von Eisenmeteoriten nach der Rhenium-Osmium-Methode. Z. Natur-forsch. 16a, 1053 (1961).

HERRMANN, R.: Flammenphotometrie. Berlin, Göttingen, Heidelberg: Springer 1956.

—, u. C. TH. J. ALKEMADE: Flammenphotometrie, 2. Auflage. Berlin, Göttingen, Heidelberg: Springer 1960.

HERZ, N., P. M. HURLEY, W. H. PINSON, and H. W. FAIRBAIRN: Age measurements from a part of the Brazilian Shield. Bull. Geol. Soc. Am. 72, 1111 (1961).

HERZOG, L. F.: Rb-Sr and K-Ca analysis and ages. Proc. 2nd conf. on nuclear proc. in geologic settings, 19, 114 (1956).

—, W. H. PINSON: Rb-Sr-age, elemental and isotopic abundance studies of stony meteorites. Am. J. Sci. 254, 555 (1956).

— —, R. F. CORMIER: Sediment age determination by Rb/Sr analysis of glauconite. Bull. Am. Ass. Petrol. Geol. 42, 717 (1958).

HESS, V. F., and J. D. ROLL: The identification of the surplus gamma-radiation from granite. Phys. Rev. 73, 916 (1948).

HESS, D. C., and R. R. MARSHALL: The isotopic compositions and concentrations of lead in some chondritic stone meteorites. Geochim. Cosmochim. Acta 70, 284 (1960).

HEYMANN, D.: Private communication (1965a).
— Cosmogenic and radiogenic helium, neon, and argon in amphoteric chondrites. J. Geophys. Res. **70**, 3735 (1965b).
— M. E. LIPSCHUTZ, B. NIELSEN, and E. ANDERS: Canyon Diablo meteorite: Metallographic and mass spectrometric study of 56 fragments. J. Geophys. Res. (1966), in press.
HILLER, J. E.: Die mineralischen Rohstoffe. Stuttgart: E. Schweizerbart'sche Verlagsbuchhandlung 1962.
HINTENBERGER, H.: High sensitivity mass spectroscopy in nuclear studies. Ann. Rev. Nucl. Sci. **12**, 435 (1962).
— H. KOENIG and H. WÄNKE: Über den Helium- und Neongehalt von Steinmeteoriten und deren radiogene und kosmogene Alter. Z. Naturforsch. **17a**, 1092 (1962).
— Massenspektroskopie kleinster Substanzmengen in der Kernphysik und in der Isotopenkosmologie. I. Naturwissenschaften **51**, 473 (1964a).
— Massenspektroskopie kleinster Substanzmengen in der Kernphysik und in der Isotopenkosmologie. II. Naturwissenschaften **51**, 497 (1964b).
— H. KÖNIG, L. SCHULTZ and H. WÄNKE: Radiogene, spallogene und primordiale Edelgase in Steinmeteoriten. Z. Naturforsch. **19a**, 327 (1964).
— — — — Radiogene, spallogene und primordiale Edelgase in Steinmeteoriten III. Z. Naturforsch. **20a**, 983 (1965).
HIRZEL, O., and H. WÄFFLER: On the radioactivity of K^{40}. Phys. Rev. **74**, 1553 (1948).
— — Über die beim Zerfall des K^{40} emittierte γ-Strahlung. Helv. Phys. Acta **19**, 216 (1946).
HOEGFELDT, E.: Ion-exchange equilibria II, Activities of the components in ion exchangers. Arkiv Kemi **5**, 147 (1952).
HOLMES, A.: Age of the base of the Cambrian. Nature (Lond.) **187**, 35 (1960a).
— A revised geological time scale. Edingburgh Geol. Soc. Trans. **17**, 183 (1960b).
HONDA, M., and J. R. ARNOLD: Effects of cosmic rays on meteorites. Science **143**, 203 (1964).
HOUTERMANNS, F. G., O. HAXEL u. J. HEINTZE: Die Halbwertzeit des K^{40}. Z. Physik **128**, 657 (1950).
HOWER, J., P. M. HURLEY, W. H. PINSON, and H. W. FAIRBAIRN: The dependence of K-Ar age on the mineralogy of various particle size ranges in a shale. Geochim. Cosmochim. Acta **27**, 405 (1963).
HUNT, G.: Time of Purcell eruption in southeastern British Columbia and Southwestern Alberta. Alberta Soc. Petrol. Geol. J. **10**, 438 (1962).
HUNTLEY, H. E.: Production of helium by cosmic rays. Nature (Lond.) **161**, 356 (1948).
HURLEY, P. M.: Age study of lower Paleozoic glauconites. M.I.T. Dept. Geol. and Geophys., U.S.A.E.C. 6th Ann. Rept. NYO-3939, 53 (1958a).
— A/K age of a Pacific Ocean core sample. M.I.T. Dept. Geol. and Geophys., U.S.A.E.C. 6th Ann. Rept., NYO-3939, 114 (1958b).
— H. W. FAIRBAIRN, W. H. PINSON, and J. W. WINCHESTER: Variations in isotopic abundance of strontium, calcium, and argon and related topics. Sixth Annual Prog. Report for 1958 to U.S.A.E.C. NYO-3939, 28 (1958).
— J. HOWER, and R. F. CORMIER: Reliability of glauconite for age measurement by the K-Ar and Rb-Sr methods. M.I.T. Dept. Geol. and Geophys., U.S.A.E.C. 7th Ann. Rept., NYO-3940, 23 (1959a).
— S. R. HART, W. H. PINSON, and H. W. FAIRBAIRN: Authigenic versus detrital illite in sediments (Abstract). Bull. Geol. Soc. Am. **70**, 1622 (1959b).

HURLEY, P. M.: Study of the age of K-phases in deep ocean sediments. M.I.T., Dept. Geol. and Geophys., U.S.A.E.C. 8th Ann. Rept., NYO-3941, 267 (1960).
— R. F. CORMIER, J. HOWER, H. W. FAIRBAIRN, and W. H. PINSON JR.: Reliability of glauconite for age measurement by K-Ar and Rb-Sr methods. Bull. Am. Ass. Petrol. Geol. 44, 1793 (1960a).
— H. W. FAIRBAIRN, W. H. PINSON, and G. FAURE: K-A and Rb-Sr minimum ages for the Pennsylvanian section in the Narragansett Basin. Geochim. Acta 18, 247 (1960b).
— Glauconite as a possible means of measuring the age of sediments in geochronology of rock systems. Ann. N. Y. Acad. Sci. 91, 294 (1961).
— D. G. BROOKINS, W. H. PINSON, S. R. HART, and H. W. FAIRBAIRN: K-Ar age studies of Mississippi and other river sediments. Bull. Geol. Am. 72, 1807 (1961a).
— N. H. FISHER, W. H. PINSON, and H. W. FAIRBAIRN: Geochronology of Proterozoic granites in northern territory, Australia. Part I K-Ar and Rb-Sr dates. Bull. Geol. Soc. Am. 72, 653 (1961b).
— H. W. FAIRBAIRN, W. H. PINSON JR., and J. HOWER: Unmetamorphosed minerals in the Gunflint Formation used to test the age of the Animikie. J. Geol. 70, 489 (1962a).
— H. HUGHES, W. H. PINSON JR., and H. W. FAIRBAIRN: Radiogenic argon and strontium diffusion parameters in biotite at low temperatures obtained from Alpine Fault uplift in New Zealand. Geochim. Cosmochim. Acta 26, 67 (1962b).
— J. M. HUNT, W. H. PINSON, and H. W. FAIRBAIRN: K-Ar age values on the clay fractions in dated shales. Geochim. Cosmochim. Acta 27, 279 (1963a).
— B. C. HEEZEN, W. H. PINSON, and H. W. FAIRBAIRN: K-Ar age values in pegalic sediments of the North Atlantic. Geochim. Cosmochim. Acta 27, 393 (1963b).
INGHRAM, M. G., H. BROWN, C. PATTERSON, and D. C. HESS: The branching ratio of K^{40} radioactive decay. Phys. Rev. 80, 916 (1950).
— Trace element determination by the mass spectrometer. J. Phys. Chem. 57, 809 (1953).
— D. C. HESS, and R. J. HAYDEN: Mass spectroscopy. Phys. Res. Nat. Bur. Stand. Circular 522, 257 (1953).
— Stable isotope dilution analysis. Ann. Rev. Nucl. Sci. 4, 81 (1954).
—, and R. J. HAYDEN: Handbook on mass spectroscopy. Nucl. Sci. Series Rept. 14, Nat. Res. Council 1954.
JAEGER, E., and H. FAUL: Age measurements on some granites and gneisses from the Alps. Bull. Geol. Soc. Am. 70, 1553 (1959).
— — Altersbestimmungen an einigen Schweizer Gesteinen und dem Granit von Baveno. Schweiz. Min. Petr. Mitt. 40, 10 (1960).
— J. GEISS, E. NIGGLI, A. STRECKEISEN, E. WENK u. H. WÜTHRICH: Rb-Sr Alter an Gesteinsglimmern der Schweizer Alpen. Schweiz. Min. Petr. Mitt. 41, 255 (1961).
— Rb-Sr age determination on micas and total rocks from the Alps. J. Geophys. Res. 67, 5293 (1962).
— E. NIGGLI, and H. BAETHGE: Two standard minerals, biotite and muscovite, for Rb-Sr and K-Ar age determinations, samples Bern 4B and Bern 4M from a gneiss from Brione, Valle Verzasca (Switzerland). Schweiz. Min. Petr. Mitt. 43, 465 (1963).
JANOSCHEK, R.: Das Alter der Moldavitschotter in Mähren. Anzeiger der Akad. d. Wiss. in Wien, Math.-naturwiss. Klasse 71, (17), 195 (1934).
JOHNSON, M. R. W.: Some time relations of movement and metamorphism in the Scottish Highlands. Geol. en Mijnbouw 42, 121 (1963).

KAISER, W.: Unpublished work (1965).

—, u. J. ZÄHRINGER: Kalium-Analysen von Amphoterit-Chondriten und deren K-A-Alter. Z. Naturforsch. 20 a, 963 (1965).

KALBITZER, S.: Experimente zur Edelgasdiffusion in Alkalihalogenid-Einkristallen. Z. Naturforsch. 17 a, 1071 (1962).

KAPLAN, G., et F. LEUTWEIN: Contribution a l'etude geochronologique du massif granitique de Vire (Normandie). C. R. Acad. Sci. (Paris) 256, 2006 (1963).

KARPINSKAYA, T. B.: Synthetic introduction of argon into mica at high pressures and temperatures. Akad. Nauk USSR Izv. Ser. Geol., no. 8, 87 (1961).

— Synthesis of argon-bearing muscovite. Akad. Nauk USSR Izv. Ser. Geol. no. 11, 95 (1964).

KATCOFF, S., O. A. SCHAEFFER, and J. M. HASTINGS: Half-Life of I^{129} and the age of the elements. Phys. Rev. 82, 688 (1951).

KAWANO, Y., and Y. UEDA: K-Ar-Dating on the Igneous Rocks in Japan (I). The Science Reports of the Tohoku University Series III, V. IX, no. 1 (1964).

KAZAKOV, G. A., and N. I. POLEVAYA: Some preliminary data on the development of a post-Precambrian scale of absolute geochronology based on glauconites. Geokhimiya 4, 296 (1958).

— —, and G. A. MURINA: The absolute age of Precambrian sedimentary strata of the Russian Platform and the Urals. Trans. Interdepartmental Stratigraphic Conference 1960.

KEIL, K.: Fortschritte in der Meteoritenkunde. Fortschr. Mineral. 38, 202 (1960).

— Possible correlation between classifications and potassium-argon ages of chondrites. Nature (Lond.) 203, 511 (1964).

KEMPE, W., and J. ZÄHRINGER: preprint (1965).

KENDALL, B. R. F.: Isotopic composition of potassium. Nature (Lond.) 186, 225 (1960).

KHUTSAIDZE, A. L.: Experimental investigation of diffusion of radiogenic argon in feldspars and micas. Geochemistry 11, 1063 (1962).

KIRSTEN, T., D. KRANKOWSKY, and J. ZÄHRINGER: Edelgas- und Kalium-Bestimmungen an einer größeren Zahl von Steinmeteoriten. Geochim. Cosmochim. Acta 27, 13 (1963).

KISTLER, R. W., P. C. BATEMAN, and W. W. BRANNOCK: Isotopic ages of minerals from granitic rocks of the central Sierra Nevada and Inyo Mountains, California. Geol. Soc. Am. Bull. 76, 155 (1965).

KLEMPERER, O.: On the radioactivity of potassium and rubidium. Proc. Roy. Soc. (London) A 148, 638 (1935).

KLYAROVSKIY, V. M., A. M. DMITRIYEV, V. S. KOZHEVNIKOV, and N. KH. BELOUS: Absolute age of the Western Siberian iron ore basin according to glauconites. Akad. Nauk USSR, Commission for the Determination of the Absolute Age of Geologic Formations 9th Session, 216, 1961.

KÖNIG, H., u. H. WÄNKE: Uranbestimmung an Steinmeteoriten mittels Neutronenaktivierung über die Xenon-Isotope 133 und 135. Z. Naturforsch. 14a, 866 (1959).

KOENIGSWALD, G. H. R. VON, W. GENTNER, and H. J. LIPPOLT: Age of basalt flow at Olduvai, East Africa. Nature (Lond.) 192, 720 (1961).

KOHLHÖRSTER, W.: Gammastrahlen an Kaliumsalzen. Naturwissenschaften 16, 28 (1928).

KOHMAN, T. P.: Chronology of nucleosynthesis and extinct natural radioactivity. J. Chem. Educ. 38, 73 (1961).

KOLTHOFF, I. M., and P. J. ELVING: Treatise on Analytical Chemistry. New York, London: Interscience Publishers 1961.

KOLTZOVA, T.: The development of a rapid argon micromethod for age determinations of minerals and investigations about the reliability of its application (in Russian). Bulletin of the IV Session of the Committee on the Determination of the Geologic Formations 1955, Moscow 1957.

KOMLEV, L. V., V. G. SAVONENKOV, S. I. DANILEVICH, K. S. IVANOVA, G. N. KUCHINA, and A. D. MIKHALEVSKAYA: Geological significance of regional rejuvenation of ancient formations in the southwestern part of the Ukrainian crystalline shield. Geochemistry no. 3, 219. Transl. from Geokhimiya Publ. Akad. Sci. USSR, no. 3, 195 (1962).

KOUVO, O.: Radioactive age of some Finnish Precambrian minerals. Bull. comm. géol. Finlande 182, 1 (1958).

KRANKOWSKY, D.: Massenspektrometrische Kaliumanalysen an Steinmeteoriten nach dem Isotopenverdünnungsverfahren (Diplomarbeit). Heidelberg 1960.

—, and J. ZÄHRINGER: K-Ar ages of meteorites (this book).

KRAUS, K. A., TH. A. CARLSON, and J. S. JOHNSON: Cation-exchange properties of zirconium (IV)-tungsten (VI) precipitates. Nature (Lond.) 177, 1128 (1956).

KRINOV, E. L.: Principles of meteorites. Oxford: Pergamon Press 1960.

KRUMMENACHER, D., et J. F. EVERNDEN: Determination d'age isotopique faites quelques roches des Alpes par la methode potassium-argon. Schweiz. Min. Petr. Mitt. 40, 267 (1960).

— Determinations d'age isotopique faites sur quelques roches de l'himalaya du Nepal par la methode potassium-argon. Schweiz. Min. Petr. Mitt. 41, 273 (1961).

— C. M. MERRIHUE, R. O. PEPIN, and J. H. REYNOLDS: Meteoritic krypton and barium versus the general isotopic anomalies in meteorite xenon. Geochim. Cosmochim. Acta 26, 231 (1962).

KRUSHCHEV, D. P., and B. B. ZAIDIS: Determination of the age of the Romensk potassium salts. Geochemistry no. 12, 1211. Transl. from Geokhimiya Publ. Acad. Sci. USSR no. 12, 1154 (1963).

KRYLOV, A. YA., and YU-I. SILIN: The applications of the argon method of age determination to clastic sedimentary rocks. Akad. Nauk USSR Izv. Ser. Geol. 1, 56 (1960).

— The possibility of using the absolute age of metamorphic and fragmental rocks in paleogeography and paleotectonics. Ann. N. Y. Acad. Sci. 91, 324 (1961).

— YU. I. SILIN, L. YA. ASTRASHENOK, and A. V. LOVTSYUS: Absolute age of rocks in the Mirny Region, Antarctica. Geochemistry, no. 11, 1155. Transl. from Geokhimiya Publ. Acad. Sci. USSR no. 11, 1034 (1961a).

— A. P. LISITSYN, and YU. I. SILIN: Significance of the argon-potassium ratio in ocean sediments. Akad. Nauk USSR Izv. Ser. Geol. 3, 87 (1961b).

— P. S. VORONOV, and YU. I. SILIN: The absolute age of the eastern Antarctica crystalline basement. Dokl. Acad. Sci. USSR, Earth Sci. Sect. 143, 18 (1964). Transl. from Dokl. Akad. Nauk USSR 143, 184 (1962).

— Comparison of absolute ages of rocks, micas and feldspars carried out by the argon method. Akad. Nauk USSR Kom. Opredeleniyu Absolyut. Vozrasta Geol. Formatisy Trudy, 11th Sess., p. 194, 1963.

KULP, J. L., and H. NEUMANN: Some potassium-argon ages of rocks from the Norwegian Basement. Ann. N. Y. Acad. of Sci. 91, 469 (1960).

— L. E. LONG, C. E. GIFFIN, A. A. MILLS, R. STJ. LAMBERT, B. J. GILETTI, and R. K. WEBSTER: Potassium-argon and rubidium-strontium ages of some granites from Britain and Eire. Nature (Lond.) 185, 495 (1960).

—, and F. D. ECKELMANN: Potassium-argon isotopic ages on micas from the southern Appalachians. Ann. N. Y. Acad. Sci. 91, 408 (1961).

— Geologic time scale. Science 133, 1105 (1961).

KULP, J. L., and J. ENGELS: Discordances in K-Ar and Rb-Sr Isotopic Ages, in "Radioactive Dating", International Atomic Energy Agency, Vienna, 219, 1963.

KURONAYANAGI, T., T. TAMURA, K. TANAKA, and H. MORINAGA: Potassium — 47. Nucl. Physics 50, 417 (1964).

KUZ'MIN, A. M.: Retention of argon in microline. Geochemistry 5, 485 (1961).

LAGERWALL, T., and K. E. ZIMEN: The Kinetics of Rare-Gas Diffusion in Solids. EUR 1372.e EURAEC report no. 772 (1964).

LACROIX, A.: Les Tectites de l'Indochine. Arch. Mus. hist. nat. Paris, Ser. 6, 8, 193 (1932).

— Les tectites sans formes figurees de l'Indochine. C. R. Acad. sci. (Paris) 200, 2129 (1935a).

— Les tectites de l'Indochine et de ses abords et la Cote d'Ivoire. Arch. Mus. hist. nat. Paris, Ser. 6, 12, 151 (1935b).

LÄMMERZAHL, P., and J. ZÄHRINGER: K-Ar-Altersbestimmungen an Eisenmeteoriten. II. Spallogenes Ar^{40} und Ar^{40}-Ar^{38}-Bestrahlungsalter. Submitted to Geochim. Cosmochim. Acta (1966).

LANPHERE, M. A.: I. Geology of the Wildrose Area, Panamint Range, California. II. Geochronologic Studies in the Death Valley-Mojave Desert region. California, Ph. D. thesis, California Institute of Technology, 1962.

— G. J. WASSERBURG, A. L. ALBEE, and G. R. TILTON: Redistribution of strontium and rubidium isotopes during metamorphism, World Beater Complex, Panamint Range, California p. 269 in "Isotopic and Cosmic Chemistry", ed. by H. Craig, MILLER, S. L., and G. J. WASSERBURG: Amsterdam: North-Holland Publishing Company 1964.

—, and G. B. DALRYMPLE: P-207: An interlaboratory standard muscovite for argon and potassium analyses. J. Geophys. Res. 70, 3497 (1965).

LAPHAM, D. M., and W. A. BASSET: K-Ar dating of rocks and tectonic events in the Piedmont of southeastern Pennsylvania. Geol. Soc. Am. Bull. 75, 661 (1964).

LEECH, G. B., J. A. STOCKWELL, and R. K. WANLESS: Age determinations and geological studies. Report 4 — Geological Survey of Canada, paper 63—17 (1963).

LEUTZ, H., H. WENNINGER u. K. ZIEGLER: Die Halbwertzeit von Rb^{87}. Z. Phys. 169, 409 (1962).

LEVIN, B. Y., S. V. KOZLOVSKAIA, and A. G. STARKOVA: The average chemical composition of meteorites. Meteoritika 14, 38 (1956).

LEVSKII, L. K.: Cosmogenic isotopes in the Yardymlinskii meteorite. Geochemistry no. 4, 380. Transl. from Geokhimiya Publ. Acad. Sci. USSR no. 4, 358 (1961).

— Diffusion of helium from stony meteorites. Geochemistry 6, 556 (1963).

LI, PU, CHEN, YU-CHI, TU GON-CHZHI, A. I. TUGARINOV, S. I. ZYKOV, N. I. STUPNIKOVA, K.-G. KNORRE, N. I. POLEVAYA, and S. B. BRANDT: On the absolute age of rocks of the Chinese People's Republic. Geokhimiya 7, 570 (1960).

LIPPOLT, H. J.: Altersbestimmungen nach der K-Ar Methode bei kleinen Argon- und Kaliumkonzentrationen. Diss., Ruprecht-Karl-Universität, Heidelberg 1961.

—, and W. GENTNER: Argonbestimmungen an Kalium-Mineralien — X. Versuche der Kalium-Argon-Datierung von Fossilien. Geochim. Cosmochim. Acta 26, 1247 (1962).

— — K-Ar dating of some limestones and fluorites (examples of K-Ar ages with low Ar-concentrations) p. 239 in "Radioactive Dating", Internat. Atomic Energy Agency, Vienna, 1963.

— — and W. WIMMENAUER: Altersbestimmungen nach der Kalium-Argon-Methode an tertiären Eruptivgesteinen Südwestdeutschlands. Jahreshefte des Geologischen Landesamtes in Baden Württemberg 6, 507 (1963).

LIPSON, J. J., and J. H. REYNOLDS: Performance data for a high sensitivity mass spectrometer. Phys. Rev. 98, 283 (1954).
— K-A dating of sediments. Geochim. Cosmochim. Acta 10, 149 (1956).
— J. H. REYNOLDS, and R. E. FOLINSBEE: Potassium-Argon dating. Geochim. Cosmochim. Acta 10, 60 (1956).
— Potassium-argon-dating of sedimentary rocks. Bull. Geol. Soc. Am. 69, 137 (1958).
LONG, L. E., J. L. KULP, and F. D. ECKELMANN: Chronology of major metamorphic events in the southeastern United States. Am. J. Sci. 257, 585 (1959a).
— J. C. COBB, and J. L. KULP: Isotopic ages on some igneous and metamorphic rocks in the vicinity of New York City. Ann. N. Y. Acad. Sci. 80, 1140 (1959b).
— Isotopic age study, Dutchess County, New York. Bull. Geol. Soc. Am. 73, 997 (1962).
—, and J. L. KULP: Isotopic age study of the metamorphic history of the Manhattan and Reading prongs. Bull. Geol. Soc. Am. 73, 969 (1962).
LOVERING, J. F., and J. W. MORGAN: Uranium and thorium abundances in stony meteorites 1. The chondritic meteorites. J. Geophys. Res. 69, 1979 (1964).
LOWDON, J. A.: Age determinations by the Geological Survey of Canada, Report 1 — Isotopic Ages, Geol. Surv. Canada, paper 60—17 (1960).
— Age determinations by the Geological Survey of Canada, Report 2 — Isotopic Ages, Geol. Survey Canada, paper 61—17 (1961).
— R. K. WANLESS, C. H. STOCKWELL, and G. B. LEECH: Age determinations and geological studies. Report 4. Geol. Surv. Canada, paper 63—17 (1963a).
— — H. W. TIPPER, and R. K. WANLESS: Age determinations and geologic studies (including isotopic studies). Report 3 — Geol. Surv. Canada, paper 62—17 (1963b).
LUNDEGARDH, H.: Die quantitative Spektralanalyse der Elemente 1. Teil. Jena: Fischer 1929.
— Die quantitative Spektralanalyse der Elemente. 2. Teil. Jena: Fischer 1934.
MAGNUSON, N. H.: Age determination of Swedish Precambrian rocks. Geol. Fören. i Stockholm Förh. 82, 407 (1960).
MANUEL, O. K., and P. K. KURODA: Isotopic composition of the rare gases in the Fayetteville meteorite. J. Geophys. Res. 69, 1413 (1964).
MAPPER, D.: Radioactivation Analysis p. 297 in "Methods in Geochemistry", edited by SMALES and WAGER. New York, London: Interscience Publishers 1960.
MARMO, V.: On the potassium-argon ages of the granitic rocks. Schweiz. Min. Petr. Mitt. 40, 17 (1960).
MARSHALL, R. R.: Mass spectrometric study of the lead in carbonaceous chondrites. J. Geophys. Res. 67, 2005 (1962).
MASON, B.: Meteorites. New York, London: John Wiley Sons (1962a).
— The classification of chondritic meteorites. Am. Museum Novitates No. 2085 (1962b).
— Olivine composition in chondrites. Geochim. Cosmochim. Acta 27, 1011 (1963a).
— The carbonaceous chondrites. Space Sci. Rev. 1, 621 (1963b).
—, and H. B. WIIK: The amphoterites and meteorites of similar composition. Geochim. Cosmochim. Acta 28, 533 (1964).
MAURETTE, M., P. PELLAS, and R. M. WALKER: Etude des traces de fission fossiles dans le mica. Bull. soc. franç. minéral. et christ. 87, 6 (1964).
MAYER, J.: Über die böhmischen Gallmeyarten, die grüne Erde der Mineralogen, die Chrysoliten von Thein und die Steinart von Kuchel. Böhmische Gesell. Wiss. Abh. Jahr. 1787, 265. Prag and Dresden 1788.
MAYNE, K. J.: Mass spectrometry. Rep. Progr. Physics XV, 24 (1952).

McDOUGALL, I.: Determination of the age of a Basic igneous intrusions by the potassium-argon method. Nature (Lond.) **190**, 1184 (1961).
— Potassium-argon ages from western Oahu, Hawaii. Nature (Lond.) **197**, 892 (1963a).
— Potassium-argon ages of some rocks from Viti Levu, Fiji. Nature (Lond.) **198**, 677 (1963b).
— Potassium-argon age measurements on dolerites from Antarctica and South Africa. J. Geophys. Res. **68**, 1535 (1963c).
—, and D. H. TARLING: Dating of polarity zones in the Hawaian Islands. Nature (Lond.) **200**, 54 (1963).
— W. COMPSTON, and D. D. HAWKES: Leakage of radiogenic argon and strontium from minerals in Proterozoic dolerites from British Guayana. Nature (Lond.) **198**, 564 (1963).
— Potassium-argon ages from lavas of the Hawaiian Islands. Bull. Geol. Soc. Am. **75**, 107 (1964).
—, and D. H. GREEN: Excess radiogenic argon in pyroxenes and isotopic ages on minerals from Norwegian eclogites. Norsk. Geol. Tidsskr. **44**, 183 (1964).
McDOWELL (Editor): Mass spectrometry. New York: Mc Graw Hill Book Comp. (1963).
McKINNEY, C. R., J. M. McCREA, S. EPSTEIN, H. A. ALLEN, and H. C. UREY: Improvements in mass spectrometers for the measurement of small differences in isotope abundance ratios. Rev. Sci. Instr. **21**, 724 (1950).
McNAIR, A., R. N. GLOVER, and H. W. WILSON: The decay of potassium 40. Phil. Mag. **1**, 199 (1956).
McNUTT, R. H.: A study of strontium redistribution under controlled conditions of temperature and pressure. Twelfth. Ann. Progr. Report for 1964, US Atomic Energy Commission, Contract At (30-1)- 1381, 125 (1964).
MEGRUE, G. H., and R. W. STOENNER: Rare gas contents and K-Ar ages of stony irons and achondritic meteorites. Trans. Am. Geophys. Union **46**, 126 (1965).
MEHNERT, K. R.: Argonbestimmungen an Kaliummineralien-VI. Die geologische Entwicklung des Schwarzwald-Grundgebirges anhand absoluter Altersbestimmungen nach der K-A Methode. Geochim. Cosmochim. Acta **14**, 105 (1958).
MEINKE, W. W.: Review of fundamental developments in analysis. Nucl. Analyt. Chem. **30**, 686 (1958).
MENEISY, M. Y., and J. A. MILLER: A geochronological study of the crystalline rocks of Charnwood Forest, England. Geol. Mag. **100**, 507 (1963).
MERRIHUE, C. M., R. O. PEPIN, and J. H. REYNOLDS: Rare gases in the Chondrite Pantar. J. Geophys. Res. **67**, 2017 (1962).
— Private communication (1964).
MEYER, ST., and E. SCHWEIDLER: Radioaktivität. 2. Aufl. 531. Berlin, Leipzig: B. G. Teubner 1927.
MEYER, H. A., G. SCHWACHHEIM, and M. D. DE SOUZA SANTOS: Decay of K^{40}. Phys. Rev. **71**, 908 (1947).
MILLER, J. M., and J. HUDIS: High energy nuclear reactions. Ann. Rev. Nucl. Sci. **9**, 159 (1959).
MILLER, J. A.: Potassium-argon ages of some rocks of the South Atlantic. Nature (Lond.) **187**, 1012 (1960).
—, and D. H. GREEN: Age determinations of rocks in the Lizard area. Nature (Lond.) **192**, 1175 (1961).
— A. J. BARBER, and N. H. KEMPTON: A potassium-argon age determination from a Lewisian inlier. Nature (Lond.) **197**, 1095 (1963).

MILLER, J. A., and P. E. BROWN: The age of some carbonatite igneous activity in south-west Tanganyika. Geol. Mag. 100, 276 (1963).

—, and A. E. MUSSETT: Dating basic rocks by the potassium-argon method: the Whin sill. Geophys. J. Roy. Astron. Soc. 7, 547 (1963).

—, and P. A. MOHR: Potassium-argon measurements on the granites and some associated rocks from southwest England. Geol. J. (formerly Liverpool and Manchester Geol. J.) 4, 105 (1964).

— — Potassium-argon age determination on rocks from St. Kilda and Rockall. Scot. J. Geol. 1, 93 (1965).

MIKHEYENKO, V. I., and N. I. NENASHEV: Absolute age of formation and relative age of intrusion of the kimberlites of Yakutia. Intern. Geol. Review 4, 916 (1962).
Transl. from Trudy 9th Sess. Kom. Opr. Absol. Vozrasta Geol. Form., Nauk USSR, Moskow-Leningrad, p. 146 (1961).

MIRKINA, S. L., E. K. GERLING, and YU. A. SHUKOLYUKOV: Determination of the age of the alcalic complexes of the Middle Urals by lead-isotopic and potassium-argon methods. Geochemistry, no. 8, 745. Transl. from Geokhimiya Publ. Acad. Sci. USSR, no. 8, 643 (1962).

MOLJK, A., R. W. P. DREVER, and S. C. CURRAN: Trace-quality analysis: neutron activation applied to potassium-mineral dating. Nucleonics 13, 44 (1955).

MØLLER, CHR.: On the capture of orbital electrons by nuclei. Phys. Rev. 51, 84 (1937).

MOORBATH, S.: Radiochemical Methods, p. 247. In "Methods in Geochemistry", edited by SMALES and WAGER. New York, London: Interscience Publishers 1960.

MORINAGA, H., and G. WOLZAK: Potassium — 45. Phys. Letters 11, 148 (1964).

MOUSUF, A. K.: K⁴⁰ radioactive decay: its branching ratio and its use in geological age determinations. Phys. Rev. 88, 150 (1952).

MUELLER, R. F., and E. OLSON: Silicates in some iron meteorites. Nature (Lond.) 201, 597 (1964).

MÜLLER, O.: Unpublished work (1965).

—, and J. ZÄHRINGER: K-Ar-Altersbestimmungen an Eisenmeteoriten III. Kalium- und Argon-Bestimmungen. Preprint (1966). Submitted to Geochim. Cosmochim. (1966).

— Publication in preparation (1966).

MUEHLBERGER, W. R., R. E. DENISON, and E. G. LIDIAK: Buried basement rocks of the United States of America and Canada. Final report ARPA contract AF 49 (638)-115 (1964).

MURINA, G. A., and V. D. SPRINTSSON: On retention of radiogenic argon in glauconites. Geokhimija 5, 459 (1960). Transl. in Geochemistry 5, 489 (1961).

MURTHY, V. R., and C. C. PATTERSON: Primary isochrone of zero age for meteorites and the earth. J. Geophys. Res. 67, 1161 (1962).

MURTHY, V. R., and W. COMPSTON: Rubidium-strontium ages of chondrules and carbonaceous chondrites. Trans. Am. Geophys. Union 45, 91 (1964).

NAUGHTON, J. J.: Possible use of Ar³⁹ in the potassium argon method of age determination. Nature (Lond.) 197, 661 (1963).

NEUMANN, H.: Apparent ages of Norwegian minerals and rocks. Norsk. Geol. Tidsskr. 40, 173 (1960).

NEWMANN, F. H., and H. J. WALKE: The radioactivity of potassium and rubidium. Phil. Mag. 19, 767 (1935).

NICOLAYSEN, L.-O.: Solid diffusion in radioactive minerals and the measurement of absolute age. Geochim. Cosmochim. Acta 11, 41 (1957).

NIER, A. O.: Evidence for the existence of an isotope of potassium of mass 40. Phys. Rev. 48, 283 (1935).

NIER, A. O.: A mass-spectrographic study of the isotopes of argon, potassium, rubidium, zinc and cadmium. Phys. Rev. 50, 1041 (1936).
— A mass spectrometer for isotope and gas analysis. Rev. Sci. Inst. 18, 398 (1947).
— Isotopic analysis of trace quantities of rare gases. Advances Mass Spectrometry Proc. Conf. Univ. London, 1958, 507, 195 (1959).
— Small general purpose double focusing mass spectrometer. Rev. Sci. Instr. 31, 1127 (1960).
NODDACK, W., and G. ZEITLER: Altersbestimmungen an Graniten des Fichtelgebirges nach der Argonmethode. Z. Elektrochem. 58, 643 (1954).
— — Über die Diffusion von Argon in Kalifeldspäten. Z. Elektrochem. 60, 1192(1956).
NORTON, F. J.: Helium-Diffusion through glass. J. Am. Ceramic Soc. 36, 90 (1953).
— Permeation of gases through solids. J. Appl. Phys. 28, 34 (1957).
— Permeation of gaseous oxygen through vitreous silica. Nature (Lond.) 191, 701 (1961).
— Gas permeation through the vacuum envelope. "1961 Transactions of the eight Vacuum Symposion and second International Congress", p. 8. London: Pergamon Press 1962.
NOVIKOV, E. A.: New data on the absolute age of the Taurian shales of Crimea. Akad. Nauk USSR Dokl. 153, 1152 (1963).
Nuclear data sheets (1958—60). Edited by the Nuclear Data Group of the National Academy of Sciences-National Research Council, Washington D. C.
OBRUČEV, S. V.: The absolute age of the precambrian rocks of the USSR. Akad. Nauk USSR, Moscow-Leningrad 1965.
O'LEARY, W. J., and J. PAPISH: Methode zur Trennung von Caesium, Rubidium und Kalium mittels Luteo-phosphormolybdänsäure. Ind. Eng. Chem. Anal. Ed. 6, 107 (1934).
OKANO, J.: Quantitative analysis of radiogenic argon in micas by isotope dilution method. Mass spectroscopy (Jap. jour., engl. text) 17, 23 (1961).
O'KEEFE, J.: The origin of tektites, in "Tektites", 167. Ed. O'KEEFE, J. Chicago: University of Chicago Press 1963.
ÖPIK, E. J.: Meteor impact on solid surface. Irish Astron. J. 5, 14 (1958).
— Notes on the theory of impact craters. Proc. of the Geophysical Laboratory. Lawrence Radiation Laboratory Cratering Symposium, Washington D. C. 2, Report UCRL-6438 (1961).
OVCHINNIKOV, L. N., A. S. SHUR, and M. V. PANOVA: On the absolute age of some igneous, metamorphic, and sedimentary formations of the Urals. Akad. Nauk USSR Izv. Ser. Geol. 10, 3 (1957).
—, and M. A. HARRIS: Absolute age of the geological formations of the Urals and the pre-Urals. Rept. 21st Intern. Geol. Congr., Norden, Part 3, 33 (1960).
— A. S. SHUR, and M. V. PANOVA: Some results of application of the potassium-argon method for determination of the absolute age of minerals and rocks of the Urals. Akad. Nauk USSR, Commission for the Determination of the Absolute Age of Geologic Formations, 6th sess., p. 8 (1960).
OZIMA, M.: Some experiments in potassium-argon dating. J. Geophys. Res. 64, 2033 (1959).
PAHL, M., J. HIBY, F. SMITS, and W. GENTNER: Mass spectrometric determination of argon from potash salts. Z. Naturforsch. 5a, 404 (1950).
PANETH, F. A., and K. PETERS: Heliumuntersuchungen I. Über eine Methode zum Nachweis kleinster Heliummengen. Z. phys. Chem. 134, 353 (1928).
— P. REASBECK, and K. I. MAYNE: Helium 3 content and age of meteorites. Geochim. Cosmochim. Acta 2, 300 (1952).
— Mikroanalyse der Edelgase. Endeavour 12, 4 (1953).

PARKINS, W. E., G. J. DIENES, and F. W. BROWN: Pulse-annealing for the study of relaxation processes in solids. J. Appl. Phys. **22**, 1012 (1951).

PATTERSON, C., H. BROWN, G. TILTON, and M. G. INGHRAM: Concentration of uranium and lead and the isotopic composition of lead in meteoritic material. Phys. Rev. **92**, 1234 (1953).

— The Pb²⁰⁷/Pb²⁰⁶ ages of some stone meteorites. Geochim. Cosmochim. Acta **7**, 151 (1955).

— Age of meteorites and the earth. Geochim. Cosmochim. Acta **10**, 230 (1956).

PEARN, W. C., E. E. ANGINO, and D. STEWART: New isotopic age measurement from the McMurdo Sound area, Antarctica. Nature (Lond.) **199**, 685 (1963).

PEPIN, R. O., J. H. REYNOLDS, and F. TURNER: Shock emplaced argon in a stony meteorite. 2. A comparison with natural argon in its diffusion. J. Geophys. Res. **69**, 1406 (1964).

PERMYAKOV, V. V., and N. A. SAVCHENKO: On the relationship between determinations of absolute and relative age of the shales of the central Caucasus. Akad. Nauk USSR Dokl. **149**, 1414 (1963).

PERRY, S. H.: The metallography of meteoritic irons. U.S. Natl. Museum Bulletin 184 (1944) (Smithsonian Institution Washington D. C.).

PINSON, W. H., C. C. SCHNETZLER, and E. BEISER: Rb-Sr age studies of stone meteorites. Tenth Ann. Prog. Rept. U.S. Atomic Energy Commission Contract AT (30-1), 1381, 19 (1962).

— C. C. SCHNETZLER, E. BEISER, H. W. FAIRBAIRN, and P. M. HURLEY: Rb-Sr age of stony meteorites. Eleventh Ann. Prog. Rept. U.S. Atomic Energy Commission, Contract AT (30-1), 1381, 7 (1963).

— C. C. SCHNETZLER, E. BEISER, H. W. FAIRBAIRN, and P. M. HURLEY: Rb-Sr age of stony meteorites. Geochim. Cosmochim. Acta **29**, 455 (1964).

POHLHAUSEN, K.: Zur näherungsweisen Integration der Differentialgleichung der laminaren Reibungsschicht. Z. angew. Math. Mech. 1, 235 (1921).

POLEVAYA, N. I., and G. A. KAZAKOV: New data on the geochronology of the late Precambrian. Dokl. Acad. Sci. USSR, Earth Sci. Sect., **135**, 1113 (1961). Transl. from Dokl. Akad. Nauk USSR **135**, 162 (1960).

— The absolute geochronological scale. Akad. Nauk USSR Doklady **134**, 1173 (1960).

— G. A. MURINA, and G. A. KAZAKOV: Absolute age of early Paleozoic and late Precambrian glauconites of the European part of the USSR. Akad. Nauk USSR Dokl. **133**, 1425 (1960a).

— — V. D. SPRINTSSON, and G. A. KAZAKOV: Determination of the absolute age of sedimentary and volcanogenic formations. Internatl. Geol. Cong., 21st, Copenhagen, 1960, Dokl. Soviet. Geol, Problema **3**, 32 (1960b).

— G. A. KAZAKOV, and G. A. MURINA: Glauconites as an indicator of geologic time. Geokhimiya 1, 3 (1960c).

— Scale of absolute geochronology according to glauconite. Akad. Nauk USSR Lab. Geol. Dokembriya Trudy 12, 123 (1961a).

— Data for compilation of the post-Precambrian scale of absolute geochronology. Akad. Nauk USSR, Commission for the Determination of the Absolute Age of Geologic Formations, 9th Session, p. 173 (1961b).

— G. A. MURINA, and G. A. KAZAKOV: Utilization of glauconite in absolute dating. Ann. N. Y. Acad. Sci.: Geochronology of Rock Systems. J. L. KULP (Editor) **91**, 2 (1961).

—, and G. A. KAZENKOV: Age differentiation and correlation of ancient unfossiliferous rocks according to the ratio Ar⁴⁰/K⁴⁰ in glauconites. Akad. Nauk USSR Lab. Geol. Dokembriya Trudy 12, 103 (1961).

POLKANOV, A. A., and E. K. GERLING: K-A and Rb-Sr methods and age of Pre-cambrian of USSR. Trans. Am. Geophys. Union 39, 713 (1958).
— — The Pre-cambrian geochronology of the Baltic shield. Rept. 21st Intern. Geol. Congr., Norden, Part 9, 183 (1960).
POOLE, W. H., J. BELAND, and R. K. WANLESS: Minimum age of Middle Ordovician rocks in southern Quebec. Geol. Soc. Am. Bull. 74, 1063 (1963).
— D. G. KELLEY, and E. R. W. NEALE: Age and correlation problems in the Appalachian region of Canada. Roy. Soc. Canada Spec. Pub. 8, 61 (1964).
POLUEKTOV, N. S.: Techniques in Flame Photometric Analysis. Russian Trans-lation. New York: Noble Offset Printers 1961.
PRIOR, G. T.: The classification of meteorites. Mineral Mag. 19, 51 (1920).
— and M. H. HEY: Catalogue of meteorites. (Brit. Museum London) (1953).
PRZEWLOCKI, K., W. MAGDA, H. THOMAS, and H. FAUL: Age of some granitic rocks in Poland. Geochim. Cosmochim. Acta 26, 1069 (1962).
RAFF, P., and W. BROTZ: Über die Bestimmung des Kaliums als Tetraphenyl-borkalium. Z. analyt. Chem. 133, 241 (1951).
RAMA, S. N. I., S. R. HART, and E. ROEDDER: Excess radiogenic argon in fluid inclusions. J. Geophys. Res. 70, 509 (1965).
RANKAMA, K.: Progress in isotope geology, p. 365. London, New York: Wiley 1963.
RAPSON, J. E.: Age and aspects of metamorphism associated with the Ice River complex, British Columbia. Bull. Canad. Petrol. Geol. 11, 116 (1963).
REED, W. G., and A. TURKEVICH: Uranium, helium and the ages of meteorites. Nature (Lond.) 180, 594 (1957).
— K. KIGOSHI, and A. TURKEVICH: Determinations of concentrations of heavy elements in meteorites by activation analysis. Geochim. Cosmochim. Acta 20, 122 (1960).
REICHENBERG, D.: Properties of ion-exchange resins in relation to their structure — III. Kinetics of exchange. J. Am. Chem. Soc. 75, 589 (1953).
REYNOLDS, J. H.: A high sensitivity mass spectrometer. Phys. Rev. 98, 283 (1954).
— High sensitivity mass spectrometer for noble gas analysis. Rev. Sci. Instr. 27, 928 (1956a).
— K-Ar Dating: in Nuclear Processes in Geologic Settings, Natl. Acad. Sci., Natl. Res. Council, Publ. 400, 135 (1956b).
— Comparative study of argon content and argon diffusion in mica and feldspar. Geochim. Cosmochim. Acta 12, 177 (1957).
—, and J. J. LIPSON: Rare gases from Nuevo Laredo stone-meteorites. Geochim. Cosmochim. Acta 12, 330 (1957).
— I-Xe dating of meteorites. J. Geophys. Res. 65, 3843 (1960a).
— Isotopic composition of primordial xenon. Phys. Rev. Letters 4, 351 (1960b).
— Isotopic composition of xenon from enstatite chondrites. Z. Naturforsch. 15a, 1112 (1960c).
— Determination of the age of the elements. Phys. Rev. Letters 4, 8 (1960d).
— Rare gases in tektites. Geochim. Cosmochim. Acta 20, 101 (1960e)
RICE, S. A., and F. E. HARRIS: Polyelectrolyte gels and ion exchange reactions. Z. physik. Chem. 8, 207 (1956).
RICHARDS, J. R., J. A. COOPER, and A. W. WEBB: Potassium-argon ages on micas from the Precambrian region of northwestern Queensland. J. Geol. Soc. Austr. 10, 299 (1963).
—, and R. T. PIDGEON: Some age measurements on micas from Broken Hill, Australia. J. Geol. Soc. Austr. 10, 243 (1963).
RICHTER, A. H. K., and K. E. ZIMEN: Messungen der Argon-Diffusion im KCl und KBr. Z. Naturforsch. 20a, 666 (1965).

RIK, G. R., and JU. A. SHUKOLJUKOV: Isotopic composition of K in meteorites. Dokl. Akad. Nauk USSR **94**, 667 (1954).

RITTENBERG, D.: Some applications of mass spectrometric analysis to chemistry. J. Appl. Phys. **13**, 561 (1942).

ROBERTS, T., and A. O. NIER: The Ca^{40}-A^{40} mass difference and the radioactivity of K^{40}. Phys. Rev. **79**, 198 (1950).

ROBERTS, R. W., and T. A. VANDERSLICE: Ultrahigh vacuum. Englewood Chiffs N. J.: Prentice Hall Inc. 1963.

RODGERS, J.: Chronology of orogenic movements in the Appalachian region of eastern North America. (To be published.) (1966).

ROSE, G.: Beschreibung und Eintheilung der Meteorite auf Grund der Sammlung im mineralogischen Museum zu Berlin. Abhandl. Akad. Wiss. Berlin, **23** (1863).

ROSE, H. J. JR., I. ADLER, and F. J. FLANAGAN: X-Ray fluorescence analysis of the light elements in rocks and minerals. Appl. Spectroscopy **17**, 81 (1963).

ROSENBERG, P.: A method for diminishing the sticking of mercury in capillaries. Rev. Sci. Instr. **9**, 258 (1938).

— The design of an accurate McLeod-gauge. Rev. Sci. Instr. **10**, 131 (1939).

RUBINSHTEYN, M. M., B. G. CHIKVAIDZE, A. L. KHUTSAIDZE, and O. YA. GEL'MAN: On the use of glauconite for the determination of the absolute age of sedimentary rocks by the argon method. Akad. Nauk USSR Izv. Ser. Geol. **12**, 77 (1959).

— Some critical points of the post-Cryptozoic geological time scale. Ann. N. Y. Acad. Sci.: Geochronology of Rock Systems. J. L. KULP (Editor) **91**, 364 (1961).

— O. YA. GEL'MAN, I. G. GRIGORYEV, B. A. LASHKHI, E. D. UZNADZE, A. L. KHUTSAIDZE, and B. G. CHIKVAIDZE: Problem of compilation of the absolute geochronologic scale. Akad. Nauk USSR. Commission for the Determination of the Absolute Age of Geological Formations, 9th Session, p. 165 (1961).

— The duration of the jurassic period. Dokl. Acad. Sci. USSR, Earth Sci. Sect. **136**, 150 (1962); Transl. from Dokl. Akad. Nauk USSR **136**, 1432 (1961).

RUSSELL, R. D., H. A. SHILLIBEER, R. M. FARQUHAR, and A. K. MOUSUF: The branching ratio of potassium 40. Phys. Rev. **91**, 1223 (1953).

RUTHERFORD, E.: Einfluß der Temperatur auf die „Emanationen" radioaktiver Substanzen. Phys. Z. **2**, 429 (1901).

RUTSCH, W.: Die Diffusion der Edelgaskomponenten von Helium, Neon und Argon aus Meteoriten. Thesis at Universität Bern 1962.

SAILOR, V. L.: Energy levels in Ca^{41} and the mass difference between A^{40} and Ca^{40}. Phys. Rev. **75**, 1836 (1949).

SAMUELSON, O., E. SJÖSTRÖM, and S. FORSBLOM: Bestimmung der Alkalimetalle neben Magnesium, Calcium und anderen Metallen. Z. analyt.Chem. **144**, 323 (1955).

— Ion Exchange Separations in Analytical Chemistry. Stockholm: Almqvist and Wiksell; New York, London: John Wiley & Sons 1963.

SARDAROV, S. S.: New reactor for the release and purification of radiogenic argon. Izv. Akad. Nauk USSR Ser, Geol. **4**, 108 (1957a).

— The preservation state of radiogenic argon in microclines. Geokhimiya No **3**, 193 (1957b).

— An improved method of isotopic dilution for determination of the content of radiogenic argon in geological formations. Akad. Nauk USSR, Commission for the Determination of the Absolute Age of Geological Formations, 5th Sess., Trudy, 278 (1958).

— Retention of radiogenic argon in glauconites. Geokhimiya **10**, 905 (1963a).

— Preservation of radiogenic argon in glauconites. Geochemistry **10**, 937 (1963b).

SAURIN, E.: Sur quelques gisements de tectites de l'Indochine du Sud. C. R. Acad. Sci. (Paris) **200**, 246 (1935).

SAWYER, G. A., and M. L. WIEDENBECK: Gamma-ray of K⁴⁰. Phys. Rev. 76, 1535 (1949).
— — Decay of constants of K⁴⁰. Phys. Rev. 79, 490 (1950).
SCHAEFFER, O. A.: The effect of mass discrimination on isotope abundance measurements. Relative abundances of krypton isotopes. J. Chem. Phys. 18, 1681 (1950).
— An improved mass spectrometer ion source. Rev. Sci. Instr. 25, 660 (1954).
— High sensitivity mass spectrometry of the rare gases. Brookhaven. Nat. Lab. Report 581 and 4385 (1959).
—, and J. ZÄHRINGER: High sensitivity mass spectrometric measurement of stable helium and argon isotopes produced by high energy protons in iron. Phys. Rev. 113, 674 (1959).
— — Helium, neon and argon isotopes in some iron meteorites. Geochim. Cosmochim. Acta 19, 94 (1960).
— R. W. STOENNER, and W. A. BASSETT: Dating of Tertiary volcanic rocks by the potassium-argon method. N. Y. Acad. Sci. Ann. 91, 317 (1961).
— Radiochemistry of meteorites. Ann. Rev. Phys. Chem. 13, 151 (1962).
— R. W. STOENNER, and E. L. FIREMAN: Rare gas isotope contents and K-Ar ages of mineral concentrates from the Indarch meteorite. J. Geophys. Res. 70, 209 (1965).
SCHERBAKOV, D. I.: Scale of absolute geological time. India Geol. Survey Res. 91, 2 (1962).
SCHINDEWOLF, U.: Ionenaustauscher in der analytischen Chemie (Entwicklungen in den letzten Jahren). Angew. Chemie 69, 226 (1957).
SCHUHKNECHT, W.: Die Flammenspektralanalyse. Stuttgart: Ferdinand Enke Verlag 1961.
SCHUILING, R. D.: Some remarks concerning the scarcity of retrograde vs. progressive metamorphism. Geol. en Mijnbouw 42, 177 (1963).
SCHULZE, W.: Neutronenaktivierung als analytisches Hilfsmittel. Stuttgart: Ferdinand Enke 1962.
SCHUMACHER, E.: Altersbestimmung von Steinmeteoriten mit der Rubidium-Strontium-Methode. Z. Naturforsch. 11a, 206 (1956a).
— Isolierung von K, Rb, Sr, Ba und seltenen Erden aus Steinmeteoriten. Helv. Chim. Acta 39, 531 (1956b).
— Age of meteorites by the Rb⁸⁷-Sr⁸⁷ method. Proceedings of 2nd Conference on Nuclear Processes in Geologic Settings, NAS-NRC publ. 400, 90 (1956c).
— Quantitative Bestimmung von Rubidium und Strontium in Steinmeteoriten mit der massenspektrometrischen Isotopenverdünnungsmethode. Helv. chim. Acta 39, 538 (1956d).
Scott's Standard Methods of Chemical Analysis. Ed. N. H. FURMAN. Vol. 1. The Elements. Princeton-New York: D. van Nostrand Comp. 1962.
SEEGER, A.: Handbuch der Physik, Band VII, Teil I, 399 (1955).
Seventh Annual Progr. Report U. S. Atomic Energy Commission 1959. p. 224. Effects of hydrogen in argon analysis.
SHALGOSKY, H. I.: Fluorescent X-Ray Spectrography, p. 111, in: "Methods in Geochemistry". Edited by SMALES and WAGER. New York, London: Interscience Publishers 1960.
SHIBATA, K., J. A. MILLER, N. YAMADA, K. KAWATA, M. MURAYAMA, and M. KATADA: Potassium-argon ages of the Ingawa granite and Naegi granite. Bull. Geol. Surv. Japan 13, 317 (1962).
SHIELDS, R. M.: The Rb⁸⁷-Sr⁸⁶ age of stony meteorites. Twelfth Ann. Prog. Rept. U.S. Atomic Energy Commission Contract AT (30-1), 1381, 3 (1964).

SHILLIBEER, H. A., and R. D. RUSSELL: The potassium-argon method of geological age determination. Can. J. Phys. **32**, 681 (1954).
— — R. M. FARQUHAR, and E. A. JONES: Radiogenic argon measurements. Phys. Rev. **94**, 1793 (1954).
SIGNER, P., and A. O. NIER: The distribution of cosmic-ray produced rare gases in iron meteorites. J. Geophys. Res. **65**, 2947 (1960).
—, and H. E. SUESS: Rare gases in the sun, in the atmospheric and meteorites. In "Earth science and meteoritics", p. 241. (Ed. by GEISS, J., and E. D. GOLDBERG) Amsterdam: North Holland Publ. Co. 1963
SLEPNEV, YU., and L. L. SHANIN: The absolute age of the rare metal pegmatites of eastern Sayan. Geochemistry **1**, 67 (1961).
SMALLER, B., J. MAY, and M. FREEDMAN: Scintillation studies of potassium iodide. Phys. Rev. **79**, 940 (1950).
SMALES, A. A.: Radioactivation Analysis. Ann. Rep. Chem. Soc. London **46**, 285 (1949).
— The scope of radioactivation analysis. Atomics **4**, 55 (1953).
— Recent advance in radioactivation analysis. Proc. Int. Conf. Peaceful Uses Atomic Energy U. N. Geneva **15**, 73 (1956).
— Neutron — activation analysis, in: "Trace Analysis", p. 518. Ed. by J. H. YOE and H. J. KOCH. New York: John Wiley and Sons 1957.
—, and L. R. WAGER: Methods in Geochemistry. New York: Interscience Publishers Inc. 1960.
SMITH, L. G.: Magnetic electron multipliers for detection of positive ions. Rev. Sci. Instr. **22**, 166 (1951).
SMITH, D. G. W., H. BAADSGAARD, R. E. FOLINSBEE, and J. LIPSON: K-Ar age of Lower Devonian bentonites of Gaspe, Quebec, Canada. Bull. Geol. Soc. Am. **72**, 171 (1961).
SMITS, F., and W. GENTNER: Geochim. Cosmochim. Acta **1**, 22 (1950).
— — Argonbestimmungen an Kalium-Mineralien. I. Bestimmungen an tertiären Kalisalzen. Geochim. Cosmochim. Acta **1**, 22 (1950).
—, and J. ZÄHRINGER: Ein Massenspektrometer für kleinste Gasmengen. Z. angew. Phys. **7**, 313 (1955).
SMYTHE, W. R., and A. HEMMENDINGER: The radioactive isotope of potassium. Phys. Rev. **51**, 178 (1937).
SOROIU, M., M. CEREI, M. ONESCU, A. DANIS, and C. MANTESCU: Radiogenic argon determination by neutron activation analysis. Geochim. Cosmochim. **29**, 551 (1965).
SPENCER, L. J.: Origin of Tektites. Nature (Lond.) **132**, 571 (1933).
SPIERS, F. W.: Radioactivity of potassium. Nature (Lond.) **165**, 356 (1950).
STARIK, I. E., E. V. SOBOTOVICH, G. P. LOVTSYNS, M. M. SHATS, and A. V. LOVT-SYNS: The isotopic composition of lead in iron meteorites. Dokl. Akad. Nauk USSR **128**, 688 (1959).
— — — — Lead and its isotopic composition in iron meteorites. Dokl. Akad. Nauk USSR **134**, 555 (1960).
— M. G. RAVICH, A. YA. KRYLOV, I. YU. SILIN, L. YA. ASTRASHENOK, and A. V. LOVTSYUS: New data on the absolute ages of rocks in eastern Antarctica. Dokl. Acad. Sci. USSR, Earth Sci. Sect. **134**, 956 (1961). Transl. from Dokl. Akad. Nauk USSR **134**, 1421 (1960a).
STARIK, I. E.: Nuclear Geochronology, p. 250. Istv. Akad. Nauk. USSR Moscow 1961.
STEIGER, R. H.: Dating of orogenic phases in the Central Alps by K-Ar ages of hornblende. J. Geophys. Res. **69**, 5407 (1964).

STEVENS, J. R., and H. A. SHILLIBEER: Loss of argon from minerals and rocks due to crushing. Proc. Geol. Ass. Canad. 8, 71 (1956).

STEVENS, R. D.: Experiment on whole-rock phyllite K-Ar age determination, in "Summary of activities — Office and laboratory", 1963, Canada Geol. survey paper 64, 2, 62 (1964).

STOCKWELL, C. H.: Third report on structural provinces, orogenies, and time-classification of rocks of the Canadian Precambrian Shield. Canada Geol. Survey Paper 63—17, 125 (1963).

STOENNER, R. W., and J. ZÄHRINGER: Potassium-argon age of iron meteorites. Geochim. Cosmochim. Acta 15, 40 (1958).

STOUT, R. W.: Radioactivity of K^{40}. Phys. Rev. 75, 1107 (1949).

STROMINGER, D., J. M. HOLLANDER, and G. T. SEABORG: Table of Isotopes. Rev. Mod. Phys. 30, 585 (1958).

STRUTT, R. J.: The leakage of helium from radioactive minerals. Proc. Roy. Soc. (Lond.) A 82, 166 (1909).

SUESS, F. E.: Über den Kosmischen Ursprung der Moldavite. Mineralog. Mag. 12, 311 (1898).

SUESS, H. E.: On the radioactivity of K^{40}. Phys. Rev. 73, 1209 (1948).

—, and H. C. UREY: Abundances of the elements. Rev. Mod. Phys. 28, 53 (1956).

SUTTLE, A. D., and W. F. LIBBY: Absolute assay of beta-radioactivity in thick solids (application to naturally radioactive potassium). Anal. Chem. 27, 921 (1955).

SWEET, R. C., W. RIEMAN, and J. BEUKENKAMP: Determination of alkalimetals in insoluble silicates by ion-exchange chromatography. Anal. Chem. 24, 952 (1952).

TAYLOR, T. J., and W. W. HAVENS JR: Neutron spectroscopy and neutron inter-actions in chemical analysis. (W. G. BERL, ed.) Physical Methods in Chemical Analysis, III, p. 447. New York: Academic Press 1956.

TARASOV, L. S., E. YA. GAVRILOV, and V. I. LBEDEV: Absolute ages of the Precam-brain rocks of the Anabar Shield. Geochemistry, no. 12, 1199. Transl. from Geokhimiya Publ. Acad. Sci. USSR, no. 12, 1145 (1963).

THODE, H. G., and R. L. GRAHAM: A mass spectrometer investigation of the isotopes of xenon and krypton resulting from the fission of U^{235} by thermal neutrons. Canad. J. Res. A 25, 1 (1947).

THOMAS, H. H.: Isotopic ages on coexisting hornblende, mica, and feldspar (abs.). Trans. Am. Geophys. Union 44, 110 (1963).

THOMSON, J. J.: On the emission of negative corpuscles by the alkali metals. Phil. Mag. 10, 584 (1905).

THOMPSON, S. J., and K. I. MAYNE: The ages of three stony meteorites and a granite. Geochim. Cosmochim. Acta 7, 169 (1955).

TILTON, G. R., G. W. WETHERILL, G. L. DAVIS, and L. A. HOPSON: Ages of minerals from the Baltimore Gneiss near Baltimore, Maryland. Geol. Soc. Am. Bull. 69, 1469 (1958).

— G. L. DAVIS, G. W. WETHERILL, L. T. ALDRICH, and E. JÄGER: The ages of rocks and minerals, in Annual Report of the Director of the Geophysical Labora-tory. Carnegie Inst. Washington, Yearbook 58, 170 (1959).

— G. W. WETHERILL, G. L. DAVIS, and M. N. BASS: 1000-million-year-old minerals from the eastern United States and Canada. J. Geophys. Res. 65, 4173 (1960).

—, and S. R. HART: Geochronology. Science 140, 357 (1963).

TOLLERT, H.: Analytik des Kaliums. Stuttgart: F. Enke Verlag 1962.

TRENDELENBURG, E. A.: Ultrahochvakuum. Karlsruhe: Verlag Braun 1963.

TSCHERMAK, G.: Die Meteoriten des K. K. mineralogischen Museums am 1. Oktober 1872. Min. Petrog. Mitt. 165 (1872).
— Die mikroskopische Beschaffenheit der Meteorite. Stuttgart: E. Schweizbart'sche Verlagsbuchhandlung (1885).
TUPPER, W. M., and S. R. HART: Minimum age of the Middle Silurian in New Brunswick based on K-Ar method. Bull. Geol. Soc. Am. 72, 1285 (1961)
UREY, H. C.: The abundances of the elements. Phys. Rev. 88, 248 (1952).
—, and H. CRAIG: The composition of the stone meteorites and the origin of the meteorites. Geochim. Cosmochim. Acta 4, 36 (1953).
VAND, V.: Theory of irreversible electrical resistance changes of metallic films evaporated in vacuum. Proc. Phys. Soc. 55, 222 (1943).
VAN SCHMUS, W. R.: The geochronology of the Blind River-Bruce Mines Area, Ontario, Canada. J. Geol. 73, 755 (1965).
VENKATASUBRAMANIA, V. S., and R. S. KRISHNAN: Radioactivity and geochronology of igneous and metamorphic rocks of the Precambrian area of the Indian Peninsula. Proc. Nat. Inst. Sci. India 26 A, 89 (1960).
VERHOOGEN, J.: Temperatures within the earth. Scient. 48, 134 (1960).
VINOGRADOV, A. P., and A. I. TUGARINIOV: Some supplementary determinations of absolute age (toward universal geochronology scale.) Dokl. Acad. Sci. USSR Earth Sci. Sect. 134, 917 (1961). Transl. from Dokl. Akad. Nauk USSR 134, 1158 (1960).
— L. V. KOMLEV, S. I. DANILEVICH, V. G. SAVONKO, A. I. TUGARINOV, and M. S. FILLIPPOV: Absolute geochronology of the Ukrainian Precambrian. Rept. 21st Intern. Geol. Congr. Norden, Part 9, 116 (1960).
—, and A. I. TUGARINOV: Geochronology of the Precambrian. Geokhimija 9, 723 (1961).
— — Problems of geochronology of the Precambrian in eastern Asia. Geochim. Cosmochim. Acta 26, 1283 (1962).
— —, and S. P. ZYKOV: Über das Alter der kristallinen Gesteine Zentraleuropas. Freiberger Forsch. C 124, 39 (1962).
—, J. K. ZADOROZHNY: Edelgase in Steinmeteoriten. Geochimija 7, 587 (1964).
VISTE, E., and E. ANDERS: Cosmic ray exposure history of tektites. J. Geophys. Res. 67, 2913 (1962).
VISTELIUS, A. B.: On the question of the origin of the red beds of the Cheleken Peninsula. Attempt to use the absolute age of clastic minerals for the solution of problems of lithology and paleogeography. Akad. Nauk USSR Dokl. 124, 1307 (1959).
—, and A. YA. KRYLOV: The absolute age of the detrital part of the arenaceous-siltstone deposits of the southwest of Central Asia. Dokl. Acad. Sci. USSR, Earth Sci. Sect. 138, 516 (1962). Transl. from Dokl. Akad. Nauk USSR 138, 422 (1961).
— Paleogeographic reconstructions by absolute age determinations of sand particles. J. Geol. 72, 483 (1964).
VOTAKH, O. A., and N. D. DMITRIYEV: Correlation of the Precambrian formations of the Igarka and Turukhansk areas on the basis of absolute age data. Akad. Nauk USSR Sibirskoye Otdeleniye. Geologiya i Geofizika 7, 82 (1963).
VOYTKEVICH, G. V., and L. K. ANOKHINA: Ages of some rock complexes in the Krivoi Rog iron region. Geochemistry no. 2, 212, Transl. from Geokhimija Publ. Acad. Sci. USSR no. 2, 185 (1961).
— Isotope dilution analysis. Advances in mass spectrometry. New York: Pergamon Press 1959.

WÄNKE, H., and H. KÖNIG: Eine neue Methode zur Kalium-Argon-Altersbestimmung und ihre Anwendung auf Steinmeteorite. Z. Naturforsch. 14a, 860 (1959).
— Über den Kaliumgehalt der Chondrite, Achondrite und Siderite. Z. Naturforsch. 16a, 127 (1961).
WALDRON, J. D. (Editor): Advances in Mass Spectrometry. New York: Pergamon Press 1959.
WANLESS, R. K., and J. A. LOWDON: Isotopic age measurements on coeval minerals and mineral pairs. Canad. Geol. Survey Paper 61—17, 119 (1961).
— R. D. STEVENS, G. R. LACHANGE, and R. Y. RIMSAITE: Age determinations and geological studies. Part 1 — Isotopic ages, Report 5, Geological Survey of Canada, paper 64—17, 1 (1965).
WASSERBURG, G. J.: Argon⁴⁰: Potassium⁴⁰ dating, in "Nuclear Geology". Ed. H. FAUL. New York: J. Wiley, and Sons, Inc. 1954.
—, and R. J. HAYDEN: A⁴⁰/K⁴⁰ age determinations. Trans. Am. Geophys. Union 35, 381 (1954a).
— — The branching ratio of K⁴⁰. Phys. Rev. 93, 645 (1954b).
—, and R. J. HAYDEN: Age of meteorites by the A⁴⁰-K⁴⁰ method. Phys. Rev. 97, 86 (1955b).
— — A⁴⁰-K⁴⁰ dating. Geochim. Cosmochim. Acta 7, 51 (1955a).
— — A⁴⁰-K⁴⁰ dating. Nucl. sci. Series, Rep. No. 19, 131 (1956).
—, and K. J. JENSEN: A⁴⁰-K⁴⁰ dating of igneous rocks and sediments. Geochim. Cosmochim. Acta 10, 153 (1956).
— F. J. PETTIJOHN, and J. LIPSON: A⁴⁰-K⁴⁰ ages of micas and feldspars from the Glenarm series near Baltimore. Md., Sci. 126, 355 (1957).
—, and R. BIERI: The A³⁸ content of two potassium minerals. Geochim. Cosmochim. Acta 15, 157 (1959).
— G. W. WETHERILL, L. T. SILVER, and P. T. FLAWN: A study of the ages of the Precambrian of Texas. J. Geophys. Res. 67, 4021 (1962).
— H. CRAIG, H. W. MENARD, A. E. J. ENGEL, and C. G. ENGEL: Age and composition of a Bounty Islands granite and age of a Seychelles Islands granite. J. Geol. 71, 785 (1963).
WEBB, A. W., I. McDOUGALL, and J. A. COOPER: Retention of radiogenic argon in glauconites from Proterozoic sediments, Northern Territory, Australia. Nature (Lond.) 199, 270 (1963).
WEBSTER, R. K., J. W. MORGAN, and A. A. SMALES: Some recent Harwell analytical work on geochronology. Trans. Am. Geophys. Union 38, 543 (1957).
— Neutron activation for age determinations. Summer course on nuclear geology. Varenna 1960.
WEFELMEIER, cf. F. G. HOUTERMANS: Das Alter des Urans. Z. Naturforsch. 2a, 322 (1947).
WEIZSÄCKER, VON C. F.: Über die Möglichkeit eines dualen Beta-Zerfalls von Kalium. Phys. Z. 38, 623 (1937).
WETHERILL, G. W.: Spontaneous fission yields from uranium and thorium. Phys. Rev. 92, 907 (1953).
— L. T. ALDRICH, G. L. DAVIS: Potassium-argon ages of lepidolites (abs.). Phys. Rev. 93, 250 (1954).
— — — A⁴⁰/K⁴⁰ ratios of feldspars and micas from the same rock. Geochim. Cosmochim. Acta 8, 171 (1955).
— Discordant uranium-lead ages I. Trans. Am. Geophys. Union 37, 320 (1956).
— G. R. TILTON, G. L. DAVIS, and L. T. ALDRICH: New determinations of the age of the Bob Ingersoll pegmatite, Keystone, South Dakota. Geochim. Cosmochim. Acta 9, 292 (1956a).

WETHERILL, G. W., G. R. TILTON, G. L. DAVIS, and L. T. ALDRICH: Evaluation of Mineral Age Measurements II. Nucl. Sci. Series, Rep. Nr. 19, 151 (1956b).
— L. T. ALDRICH, and G. R. TILTON: Comparisons of K-A ages with concordant U-Pb ages of pegmatites (abs.). Trans. Am. Geophys. Union 37, 362 (1956c).
— G. J. WASSERBURG, L. T. ALDRICH, G. R. TILTON, and R. J. HAYDEN: Decay of constants of K⁴⁰ as determined by the radiogenic argon content of potassium minerals. Phys. Rev. 103, 987 (1956d).
— Radioactivity of potassium and geologic time. Science 126, 545 (1957).
— Age of the base of the Cambrian. Nature (Lond.) 187, 34 (1960).
— G. L. DAVIS, and G. R. TILTON: Age measurements on minerals from the Cutler batholith, Cutler, Ontario. J. Geophys. Res. 65, 2461 (1960).
— O. KOUVO, G. R. TILTON, and P. W. GAST: Age measurements on rocks from the Finnish Precambrian. J. Geol. 70, 74 (1962).
—, and M. E. BICKFORD: Primary and metamorphic Rb-Sr chronology in central Colorado. J. Geophys. Res. 70, 4669 (1965).
— G. R. TILTON, G. L. DAVIS, S. R. HART, and C. A. HOPSON: Age measurements in the Maryland piedmont. J. Geophys. Res. 71, 2139 (1966).
WHITHAM, B. T.: Use of molecular sieves in gas chromatography for the determination of the normal paraffins in petroleum fractions. Nature (Lond.) 182, 391 (1958).
WILLARD, H. H., and G. F. SMITH: The preparation and properties of magnesium perchlorate and its use as a drying agent. J. Am. Chem. Soc. 44, 2255 (1922).
— — The perchlorates of the alkali and alkaline earth metals and ammonium. Their solubility in water and other solvents. J. Am. Chem. Soc. 45, 286 (1923).
WILLIAMS, G. D., H. BAADSGAARD, and G. STEEN: Potassium-argon mineral dates from the Mannville group. Alberta Soc. Petrol. Geol. J. 10, 320 (1962).
WINCHESTER, J. W.: Radioactivation Analysis in Inorganic Geochemistry, in "Progress in Inorganic Chemistry". Vol. 2, p. 1. Ed. F. A. COTTON. New York: Interscience Publishers 1960.
WITTIG, G., and P. RAFF: Über Komplexbildung mit Triphenylbor. Ann. Chem. 573, 195 (1950).
WOOD, J. A.: "Physics and chemistry of meteorites" in: The Solar System IV (ed. by MIDDLEHURST, B. M., and G. P. KUIPER). Chicago: University of Chicago Press 1963.
WRAGE, E. G.: Argonbestimmungen an Kaliummineralien — VIII. Ein Näherungsverfahren zur Lösung von Diffusionsproblemen. Geochim. Cosmochim. Acta 26, 61 (1962).
YASHCHENKO, M. L., E. S. VARSHAVSKAYA, and I. M. GOROKHOV: Anomalous isotopic composition of strontium in minerals from metamorphic rocks. Geochemistry no. 5, 438. Transl. from Geokhimiya Publ. Acad. Sci. USSR no. 5, 420 (1961).
YAVNEL, A. A.: Issledovaniyes truktury Sikhote-Alinskogo meteorita. Dokl. Akad. Nauk USSR 60, 1381 (1948).
YELYANOV ELYYANOV, A. A., and V. M. MORALEV: New data on the age of the ultrabasic and alkali rocks of the Aldan shield. Dokl. Acad. Sci. USSR, Earth Sci. Sect. 141, 1163 (1963). Transl. from Dokl. Akad. Nauk USSR 141, 687 (1961).
YORK, D., R. M. MACINTRE, and J. GITTINS: Excess Ar⁴⁰ in sodalites (abs.). Trans. Am. Geophys. Union 46, 177 (1965).
ZARTMANN, R. E., G. J. WASSERBURG, and J. H. REYNOLDS: Helium, argon and carbon in some natural gases. J. Geophys. Res. 66, 227 (1961).
— A geochronologic study of the Lone Grove pluton from the Llano Uplift, Texas. J. Petrology 5, 359 (1964).

ZÄHRINGER, J., and E. L. FIREMAN: The A[40], K[41] and He[3] content of iron meteorites. Bull. Am. Phys. Soc. 344 (1956).
— Altersbestimmungen nach der K-Ar-Methode. Geol. Rundschau 49, 227 (1960).
—, u. W. GENTNER: Uredelgase in einigen Steinmeteoriten. Z. Naturforsch. 15 a, 600 (1960).
— Isotopie-Effekt und Häufigkeit der Edelgase in Steinmeteoriten und auf der Erde. Z. Naturforsch. 17 a, 460 (1962).
— K-Ar measurements of tektites in "Radioactive Dating". Vienna: International Atomic Energy Agency, 289 (1963a).
— Isotopes in tektites, in "Tektites", edited by O'KEEFE, J. University of Chicago Press, p. 141 (1963b).
— — Radiogenic and atmospheric argon content of tektites. Nature (Lond.) 199, 583 (1963).
— Isotopes chronology of meteorites. Ann. Rev. Astron. Astrophys. 2, 121 (1964).
— Chronology of chondrites with rare gas isotopes. Preprint 1964. (In press in Meteoritica.)
ZYKOV, S. I., A. I. IUGARINOV, I. V. BEL'KOV, and E. V. BIBIKOVA: The age of the oldest formations of the Kola Peninsula. Geokhimiya 4, 307 (1964).

Subject Index